Healing the Herds

Ohio University Press Series in Ecology and History

James L. A. Webb, Jr., Series Editor

Healing the Herds

Disease, Livestock Economies, and the Globalization of Veterinary Medicine

Edited by Karen Brown and Daniel Gilfoyle

OHIO UNIVERSITY PRESS

ATHENS

Ohio University Press, Athens, Ohio 45701
www.ohioswallow.com
© 2010 by Ohio University Press
All rights reserved

Printed in the United States of America
Ohio University Press books are printed on acid-free paper ⊗™

16 15 14 13 12 11 10   5 4 3 2 1

Library of Congress Cataloging-in-Publication Data
Healing the herds : disease, livestock economies, and the globalization of veterinary medicine /
edited by Karen Brown and Daniel Gilfoyle.
    p. ; cm. — (Ohio University Press series in ecology and history)
Includes bibliographical references and index.
ISBN 978-0-8214-1884-0 (cloth : alk. paper) — ISBN 978-0-8214-1885-7 (pbk. : alk. paper)
1. Veterinary epidemiology—History. 2. Livestock—History. 3. Globalization. I. Brown, Karen,
1964– II. Gilfoyle, Daniel, 1957– III. Series: Ohio University Press series in ecology and history.
[DNLM: 1. Animal Husbandry—history. 2. Disease Outbreaks—veterinary. 3. History, 18th
Century. 4. History, 19th Century. 5. History, 20th Century. 6. Veterinary Medicine—history.
SF 615 H434 2009]
SF780.9.H43 2009
338.1'76—dc22

                                                                    2009037813

Contents

Preface

This collection was selected from papers presented at a conference titled "Veterinary Science, Disease and Livestock Economies," which was organized by the editors and held at St Antony's College, Oxford, in June 2005. The idea for the conference originated from our project, sponsored by the Wellcome Trust, which explored the history of veterinary science at the Onderstepoort Research Laboratories in South Africa during the first half of the twentieth century. Our comparative reading revealed that veterinary medicine and its relations with society and the economy are underrepresented in the historiography. The relative dearth of historical studies on the subject seemed curious, given the importance of pastoralism as a productive activity in many countries and its relationship to food supply and to environmental change. The aim of the conference, therefore, was to begin to address this gap in the literature by calling for studies examining interconnections between livestock economies, veterinary science, disease, and the environment.

The call for papers was intended to attract scholars from a variety of disciplines, and we succeeded in bringing together historians, anthropologists, scientists, veterinarians, and economists. The material presented was historically and geographically widespread, ranging from the eighteenth century to the present day and covering America, Europe, Africa, Asia, and Australasia. A sizable percentage of the studies related to South Africa, probably reflecting the editors' contacts, and some of these have appeared in "Livestock Diseases and Veterinary Science," a special edition of the *South African Historical Journal* published in 2007. This book consists of case studies from the United States, the Caribbean, western Europe, parts of colonial Africa and Asia, and Australasia.

Acknowledgments

We thank the Wellcome Trust for sponsoring our postdoctoral work in South Africa and for providing funds to host the conference in which the present collection has its origins. Thanks go to William Beinart and colleagues in the Centre for African Studies, University of Oxford, for supporting this conference and covering the costs of flights from Africa. Finally we would like to thank Deborah Nightingale for providing an interesting photograph for the front cover of this book.

Karen Brown and Daniel Gilfoyle

Introduction

Karen Brown and Daniel Gilfoyle

THE PUBLICATION of a volume on livestock economies and veterinary medicine is perhaps particularly timely at the beginning of the twenty-first century, given that the interest of the urban population in animal health and welfare, at least in the West, has probably never been greater. Popular movements reflect a widespread concern about such things as animal rights, experimentation, hunting, industrial-style food production, and the threat of species extinction through exploitation and environmental change. Furthermore, certain events over the last twenty years have highlighted problems of animal diseases and their control. Foot-and-mouth disease was epizootic in Great Britain and the Netherlands during 2001, and apocalyptic images of slaughter and cremation were broadcast across the media, with considerable emotional impact. They seemed to negate modern science, with its vaccines and therapeutics, harking back to a more primitive age.

During the early 1990s, the fact that dangerous diseases may pass between animals and humans was again brought to the public consciousness by the discovery of a link between bovine spongiform encephalopathy (BSE or mad cow disease) and Creutzfeldt-Jakob disease (CJD). Presently, veterinary and medical authorities in Europe and elsewhere are concerned with the dangers posed by avian influenza, which emerged in Southeast Asia and appears to be moving westward. The disease threatens the poultry industry, but more important, from the point of view of those not involved in that economic sector, is the fear that the virus will mutate to become transmissible between humans. Fevered comparisons have been drawn in the media with the deadly "Spanish flu" epidemic of the late 1910s. While such developments offer considerable scope for sensationalist reporting, they are obviously of great importance to contemporary societies. They also raise questions about how livestock diseases have been managed in different social, political, and economic contexts.

The historical literature on the management and control of livestock diseases has, to date, largely been restricted to studies with a national or local focus. Much of what has been written so far about veterinary medicine and veterinary interventions has referred to western Europe, the United States, and South Africa, where historians have been particularly interested in examining the late nineteenth-century professionalizaton of veterinary science within the context of expanding state bureaucracies.[1] In addition, for Great Britain and the United States, there have been articles on public health issues, especially bovine tuberculosis and tapeworm infestation, which can be transmitted to humans through contaminated milk and meat, respectively. Beginning in the late nineteenth century, both governments introduced regulations dealing with food production and processing.[2] Historians have also taken an interest in contemporary diseases such as BSE and foot-and-mouth,[3] as well as infections that have historically caused devastating losses, most notably the cattle diseases contagious bovine pleuropneumonia and rinderpest.[4] In addition, there are studies linking the history of animal diseases and their control to environmental history. In the West, older ideas that livestock diseases were caused by "miasmas" or unhealthy vapors pervaded well into the twentieth century and were not automatically superseded by the reductionist germ theories of the late nineteenth century.[5] In some regions, biting arthropods such as ticks and tsetse flies transmitted specific diseases, suggesting the importance of environmental factors in their epidemiology and control. Scientists and indigenous pastoralists knew that, in some cases, wild animals played a role in the maintenance of infection, while certain plant species were toxic to domestic animals.[6] This emphasis on the ecology of disease is particularly a feature of studies on Africa, where trypanosomosis (spread by tsetse flies) has been such an important determinant of pastoral production and practices.[7]

While the historiography of veterinary medicine and animal diseases has grown considerably in recent years, relevant studies are, given the importance of the topic, still relatively few. This book is intended to assemble accounts from different parts of the world, thus providing a starting point for further comparative inquiry. Broadly speaking, four interrelated themes emerge from these chapters. Several chapters deal with the institutionalization of veterinary medicine and the role veterinary institutions came to play in state building and regulation in both metropolitan and colonial settings (in particular, those by Peter Koolmees, Ann Greene, Abigail Woods, Dominik Hünniger, Martine Barwegen, Daniel Doeppers, Rita Pemberton, Robert John Perrins, Saverio Krätli, and David Anderson). From the nineteenth century on, the professionalization of veterinary medicine was supported by improvements in the understanding of disease etiologies and the efficacy of treatments. Second, the expansion of global trade and of European colonialism was a means of disseminating Old-World pathogens to different parts of the globe, causing major cattle epizootics around the world during the second half of the nineteenth century. Rinderpest was a major problem, as the chapters by Barwegen and Doeppers reveal. Governments had little choice but to respond, so the epizootics of the late nineteenth century were an important stimulus for the establishment of state veterinary services outside Europe and America. A third theme concerns other consequences of the transfer of domestic animals and commercial pastoralism to unfamiliar environments, where livestock became susceptible to new sources of infections, such as scab and footrot in sheep, dealt with here in the Australasian context by John Fisher and Robert Peden, respectively. This gave rise to different forms of scientific study, as did exposure to tropical diseases, which contributed to the development of tropical veterinary medicine, and studies into diseases such as surra (a form of trypanosomosis) in horses and camels, which is explored here in William Clarence-Smith's chapter. Finally, several presentations illustrate the close relationship between colonialism and veterinary medicine. In some colonies, veterinary medicine was used by the state to foster the development of settler economies, and veterinary administrations became an important component of state bureaucracies (see the chapters by Fisher, Perrins, Peden, and Anderson). In colonies of conquest, however, veterinary medicine emerges as a means by which colonial administrators sought to exert control over indigenous populations, sometimes with damaging consequences for local pastoral economies. This was evident in the cases of Kenya and Niger, covered by Lotte Hughes, Saverio Krätli, and David Anderson. The book is roughly organized around these themes, though there are, of course, many overlaps.

Turning to the first of the four themes, the professionalization of veterinary medicine, Joanna Swabe has demonstrated how the nineteenth century, particularly the latter part, was a key period for the rise of the modern veterinary regime, that is, "the social practices and institutionalized behaviours that have emerged in response to the problem of maintaining animal resources and protecting human health and economy."[8] During the mid-1860s, the rinderpest epizootic in western Europe caused considerable damage among cattle in Great Britain and the Netherlands, though it was contained by more efficient systems of control in France. Rinderpest revealed the vulnerability of animal economies to infection carried through trade and the fragility of food supplies in an era of industrialization and urbanization. The control and prevention of contagious animal diseases increasingly became a priority of the state and a state function, as veterinary officials were incorporated into government bureaucracies. In Europe, strategies for containing diseases were internationalized through veterinary conferences beginning in the 1860s. Attempts to coordinate disease control across international boundaries culminated in the establishment of the Office Internationale des Épizooties in 1924, in response to the spread of foot-and-mouth disease in Europe. The increasing authority of the veterinary regime was underpinned by the professionalization of veterinary medicine, as educational standards for professional membership based on courses offered in veterinary schools were established in various countries in Europe, the United States, and South Africa.[9]

If the Americas were spared the major Old-World epizootic of the late nineteenth century—rinderpest—similar developments in veterinary medicine occurred there as administrators sought to harness science to agricultural development. In the United States, the founding of agricultural experiment stations following the 1887 Hatch Act was part of this expanding bureaucratic process.[10] A new form of applied science, economic entomology, emerged from the experiment stations where entomologists tried to eliminate pests that harmed the economy by conveying diseases. This included research into ticks, which, as many American stockowners suspected and scientists in the early 1890s proved, transmitted the cattle disease known as Texas fever (*Babesia bigemina* and *Babesia bovis*).[11] This discovery paved the way for investigations into tropical animal diseases in many parts of the world.[12] The late nineteenth century saw the establishment of state veterinary departments in British colonies, including India, South Africa, Australia, and the West Indies. Given the economic importance of pastoralism and the relative underdevelopment of the state in many colonies, the evidence suggests that while colonial veterinary services might have been initially small and frequently ineffective, they nevertheless constituted a significant part of the state-building process.

The emergence of veterinary bureaucracies during the late nineteenth century was a response to official attempts to increase the efficiency of states' administrations and facilitate economic development in order to enhance their international influence and power. In this cultural environment, supporters of the scientific enterprise developed their own rhetoric of modernity and progress. The terminology might have varied from place to place, but the American mantra of "national efficiency" advocated by scientific, economic, and conservationist lobbyists during the Progressive Era of the early twentieth century—and the concurrent ideology of constructive imperialism proposed by the British colonial secretary Joseph Chamberlain (1895–1903)—resonated with wider political ideas about development in the West, as well as in the European and Japanese settler colonies.[13]

From the late nineteenth century, various aspects of the veterinary regime were supported by increasingly sophisticated understandings of disease etiologies based on germ theory and the so-called laboratory revolution in medicine. During much of the nineteenth century, states sought to contain disease through a mixture of regulations such as quarantines to prevent the importation of sick and infectious livestock from abroad, as well as internal restrictions on stock movements and compulsory slaughter-out policies. The structures needed to enforce such measures, even at a local level, required an expansion in official personnel and increasingly, with the development of microbiological sciences, investment in immunological research as well as the creation of field veterinary departments. From the early 1880s, significant discoveries in human and animal medicine, emanating from the Louis Pasteur Institute in Paris and Robert Koch's Institut für Infectionskrankheiten in Berlin, offered new opportunities for disease control, which helped to validate the role veterinary science could play in ameliorating pastoral production.[14] Working in competition with each other, teams of scientists from both institutions began to release specific prophylactics and therapies for several diseases including anthrax, rabies, and tetanus. The search for specific preventatives accelerated in subsequent decades so that vaccines against an increasing range of animal diseases became available by the mid-twentieth century. Nevertheless, continuity with the earlier period needs to be emphasized. Stockowners practiced prophylactic inoculation before the laboratory revolution. More significantly, the older methods of control and prophylaxis—namely, import controls, quarantines, and slaughter—remain key elements of veterinary public policy right up to the present day.

Veterinary regulations and public policy are important themes in this collection, and several chapters throw further light on these issues. Peter

Koolmees takes a long-term view in his exploration of responses to epizootic diseases in the Netherlands since the eighteenth century. He demonstrates that while public responses have changed greatly in recent years, there have been strong continuities in preventive policy with a much earlier period. At the beginning of the twenty-first century, the veterinary administration continues to rely on the slaughter of infected animals as an essential preventive measure. He suggests, however, that public opinion, marked by a growing concern about the welfare of animals, may render the use of such methods increasingly difficult or unfeasible. In contrast, Dominik Hünniger analyzes administrative efforts to control epizootic disease in eighteenth-century Schleswig-Holstein. Again, he points to the importance of quarantine, slaughter, and the control of trade as the principal methods adopted by governments and draws links with the methods used to control plague in humans. Hünniger shows that the regulation of animal diseases was an important means through which the state asserted its authority and was part of the process of state formation in the preindustrial period. Several chapters deal with the establishment of veterinary regimes in the colonies. These too are concerned with the ways in which governments tried to extend their authority through the regulation of animal disease in pursuit of economic development.

Ann Greene switches our attention from epizootic disease and agricultural development in rural areas to the urban environment through an examination of the relationship between veterinarians and their most important patient, the horse, in Pennsylvania during the late nineteenth and early twentieth centuries. Toward the end of the nineteenth century, an increasingly science-based university education enabled veterinarians to attain a professional identity that allowed them slowly but surely to discard the disparaging title of "horse doctor" or "cow leech," since their university training set them apart from those who administered "folk" cures. When the importance of the horse, which fueled the Industrial Revolution and powered transport, declined from the 1920s, the veterinary profession retained its position in towns and cities. Greene's chapter illustrates the changing role of veterinarians in urban areas during the twentieth century. The route to attaining a professional identity and an indispensable role in society was, however, by no means an uninterrupted progress. As Michael Worboys has pointed out, the long-term prospects of the average practitioner in Great Britain during the nineteenth and early twentieth centuries were not promising.[15] Government appointments were few, and the major source of income, the treatment of horses, was set to decline in the face of the automobile. In addition, in the United States and parts of Europe, many stockowners remained skeptical well into the twentieth century about the benefits of veterinary science. In her chapter, Abigail Woods argues that in Great Britain, farmers were generally reluctant to call upon the services of a veterinary surgeon unless the situation was desperate. It was only during World War II, when, in an attempt to increase livestock yields, the British government sponsored research into artificial insemination to breed larger and more productive beasts, that more and more farmers felt that veterinary science had something new and worthwhile to offer them in terms of enhancing their profits.

In some parts of the world, the institutionalization and spread of Western biomedicine and veterinary controls came not in the face of economic opportunities but in response to devastating epizootics. In recent times, the second half of the nineteenth century might be regarded with some justification as a period of panzootic disease. At midcentury, contagious bovine pleuropneumonia, an insidious disease that could assume an "occult" form, spread through trade from mainland Europe to Great Britain, North America, southern Africa, Australia, and elsewhere. It became a preoccupation of embryonic veterinary services in many parts of the world. In South Africa, this disease was known as lungsickness and was closely associated with the Xhosa cattle-killing movement, which had devastating social consequences.[16] Later, rinderpest, a deadly cattle disease that had reached western Europe from central Asia during an earlier period, spread, again through trade, to India, parts of Southeast Asia, and even to Africa.

To understand the spread of diseases such as rinderpest, epidemiological factors need to be located within a broader historical and geographical context. The nineteenth century witnessed an exponential increase in trade in livestock and animal products. In the European colonies, settlers in Australia, New Zealand, and South Africa introduced Merino sheep in order to provide wool for an expanding textile market in the northern hemisphere. Colonists in these countries, as well as in Southeast Asia and the Philippines also imported cattle to feed a growing population that was becoming increasingly urbanized. Trade in livestock also enabled diseases to spread within continents, a notable example being the southward introduction of the tick-borne cattle infection East Coast fever, which entered Southern Rhodesia (now Zimbabwe) and South Africa from East Africa in 1901.[17] Unused to exposure to pathogens from outside, indigenous livestock in the importing country were particularly prone to unfamiliar infections. In Asia and Africa, colonial warfare facilitated disease transfers as the horses and oxen that accompanied foreign armies spread alien infections and contracted and disseminated more localized maladies over a wider area. It was Italian military operations in the Horn of Africa that led to the introduction of rinderpest to East Africa from India in 1887. From there, it gradually spread throughout the continent during the 1890s, obliterating herds and in some places causing famine among communities dependent on cattle. The timing of these African outbreaks coincided with rinderpest epizootics in parts of Southeast Asia, placing it on the scale of an international panzootic. "Ecological imperialism," to use Alfred Crosby's phrase, was more than the westward transferral of germs from western Europe to the Americas.[18] Ultimately, this process became global as commercial and military networks expanded. Thus, the dispersal of different diseases did not necessarily follow a linear projection from a western metropole to the colonized states. The movement of animals within continents and between different colonial states numerically extended the centers of infection for particular diseases throughout the world.

Of all the epizootics, rinderpest has received the most attention from historians, particularly of southern Africa, who have been concerned with the way in which the epizootic threw into sharp relief political and social tensions during a period of colonial conquest and nascent industrialization.[19] While Clive Spinage's book on the subject has sketched out the trajectory of rinderpest throughout the world,[20] the chapters here by Dominik Hünniger, Dan Doeppers, and Martine Barwegen provide a welcome addition to this literature with their accounts of responses to this disease in specific locales. They enable at least the beginning of a comparative analysis of reactions to rinderpest in different societies and in different time periods. Hünniger describes attempts by the authorities to control rinderpest in eighteenth-century Schleswig-Holstein as disaster management. He shows how trade embargoes and quarantines became the mainstay of preventive policy and how these could adversely affect particular social groups, as the control of animal diseases became an important way in which the state exerted and extended its authority. Doeppers and Barwegen focus, respectively, on the late nineteenth-century rinderpest epizootics in the colonies of the Philippines and Java. Again, commerce was central to the spread of disease and, as in Great Britain, India, and southern Africa, rinderpest was a powerful stimulus for the establishment and consolidation of veterinary services. Doeppers's chapter corresponds with a period of technological advance in rinderpest prophylaxis, and he shows that if government responses were initially faltering and inadequate, they were eventually replaced by more effective policies in which vaccination played an essential part. Barwegen, on the other hand, argues that veterinary policies could be misconceived and damaging, an imposition of metropolitan methods on indigenous people under a colonial regime that ignored popular beliefs and practices. Her chapter questions a too-ready acceptance of progress in the control of animal diseases during the early twentieth century.

If imperial expansion was accompanied by the transfer of pathogens to and between the colonies, the empire was certainly capable of fighting back. Colonial farmers and others, who depended one way or another on their animals, became increasingly aware that unfamiliar environments presented unfamiliar stock diseases. Ecological limitations, therefore, hampered pastoral production and became strong impetuses for scientific investigation. As in human tropical diseases, of which malaria provided a prime example, livestock infections that were attributable to biting arthropods, infectious game, or toxic flora were intimately connected with the environment, and from the late 1890s on, their study assumed an interdisciplinary character. However, whereas (at least in the British context) "tropical human medicine" became institutionalized in the metropole at the London and Liverpool Schools of Tropical Medicine, scientists of tropical animal diseases tended to pursue their studies primarily in the colonies where the infections arose.[21] Military veterinarians were, perhaps, the pioneers of these studies, one example being the British bacteriologist David Bruce. While working in northern Zululand (South Africa) during the mid-1890s, Bruce discovered that nagana (bovine trypanosomosis) was caused by a protozoan (a trypanosome) found in the blood of game and spread to cattle and horses by the bite of the tsetse fly.[22] In the French Empire, too, as Diana K. Davis has shown, some of the earliest research into animal vaccines occurred in the colonies, with the first trials of anthrax and sheep pox inoculations taking place in Algeria in the late nineteenth and early twentieth centuries.[23] To consolidate and expand this knowledge, research institutes appeared in South America, the United States, India, and various European colonies in Africa from the late nineteenth century. Scientists generated important knowledge about diseases, and their work provides an example of a field in which colonial science ran ahead of the European metropolis.[24]

One aspect of this expansion in veterinary knowledge about diseases of the tropics is illustrated by William Clarence-Smith, whose chapter usefully corrects the assumption that trypanosomosis was purely an African disease. He shows that surra, a form of trypanosomosis that affects horses and camels, was a scourge of the Asian continent. As in the case of nagana, it was a military veterinary surgeon, Griffiths Evans, who first demonstrated a connection between a species of trypanosome and this disease in 1880. His discovery was important not only because it helped people to understand how surra spread but also because it showed that Paris and Berlin were not the only centers of groundbreaking biomedical research at that time. Evans did not discover the species of fly that conveyed the disease, nor did he develop a prophylactic; but his work was nonetheless important for the expansion of veterinary medical knowledge as it encouraged further research into protozoan diseases, which ultimately revealed that flies, as well as ticks, were capable of conveying fatal infections to livestock.

A further question relating to disease and environment concerns the impact of colonial administration and Westernized veterinary regimes upon local or indigenous knowledge and practices of disease control. In many colonies, blood-sucking arthropods played an important part in the transmission of disease. Creatures such as tsetse flies and ticks were highly visible, and the evidence suggests that indigenous peoples, and indeed colonial farmers, were often aware of their connection with disease, irrespective of germ theory and other developments in Western science. They were accordingly able to develop strategies to prevent or control infections. Accounts by earlier travelers, as well as modern studies by scientists and historians, indicate that in precolonial Africa, for example, local pastoralists learned to manage their environment and avoid areas that they knew, through observation and experience, were occupied by tsetse belts or seasonally prone to tick infestation.[25] The arrival of colonial armies and settlers, however, disrupted this process, and Africans lost control not only of their land but also of their ability to manage the disease environment.

Lotte Hughes's contribution to this book looks at African approaches to the environment and explores how the Purko Maasai recollect their experiences of being ousted from the Kenyan highlands to make way for white settlers in the opening years of the twentieth century. In retrospect, they associate the consequences of eviction with longer-term problems in protecting their cattle from diseases such as nagana and East Coast fever. In Kenya, as in many other African countries, the presence of wildlife constituted another ecological factor in producing disease. For a range of cultural and economic reasons, colonial and postcolonial governments established game reserves, many of which were unfenced and bordered grazing lands.[26] Nagana, malignant catarrhal fever, and rabies are just some of the diseases that are carried by a variety of game and threaten livestock.

For Kenya and other European colonies, a notable topic was the importance of livestock economies and the development of veterinary science. This doubtless reflects the position of the colonies in the overall imperial scheme as providers of primary products. As might be expected, the story that emerges differs to some extent between colonies in which indigenous pastoralism continued to dominate in the face of relatively small numbers of colonizing farmers and the colonies of settlement to which European farmers immigrated in large numbers. In parts of East Africa, for example, colonial administrators sought to transform indigenous pastoralism into commercial production and to promote settler farming but were faced by a range of diseases, many of which were spread by ticks. The colonial authorities tried to control these through restrictions on stock movement and compulsory insecticidal dipping. In the French colonies, French veterinarians had long been involved in trying to improve the rangeland through planned farming, as French veterinary education emphasized the importance of the environment in promoting animal health and counteracting disease.[27]

In parts of Africa, especially in the literature covering the British colonies, initiatives such as compulsory dipping and intervention in pastoral land management were frequently unpopular because they undermined customary animal husbandry and represented unwelcome incursions by an alien state into the lives of nomadic pastoralists. In the 1930s, animosity intensified in Kenya and South Africa as veterinary and soil scientists became preoccupied with the issue of erosion, which they ascribed to overgrazing, and they introduced measures to force Africans to reduce their herds. In the wider context, the impact of the American Dust Bowl had a significant influence on agricultural scientists in many parts of the world as fears of desertification and the eventual collapse of rural economies began to take hold. Attention to the carrying capacity of the land became the scientific watchword for sustainable development during the 1930s and 1940s, and destocking by persuasion or force was politically imposed in many parts of colonial Africa.[28] This resonates with themes in the history of medicine, science, and technology in the colonies more generally. If Daniel Headrick has interpreted various innovations in science and medicine as "tools of empire" that enabled colonists to conquer indigenous populations and overcome hostile environmental conditions,[29] historians have more recently been concerned with the ways in which Western medicine assisted colonial administrations in extending social control over the colonized, as science underpinned militaristic public health policies and sanitary measures.[30]

David Anderson develops some of these ideas in a chapter set, like that of Lotte Hughes, in colonial Kenya. He describes the unequal distribution of veterinary services between settler farmers and indigenous pastoralists and shows how veterinary interventions among Africans were aimed at protecting European-owned cattle from disease through the imposition of disruptive and damaging quarantines. He reminds us that veterinary medicine was by no means for the benefit of all, by illuminating how veterinary policy was skewed toward the aim of obtaining supplies for an embryonic meat-packing industry. He outlines the tensions these policies engendered and provides a critique of the myth of the "economic irrationality" of pastoral producers. From a West African perspective, Saverio Krätli examines French interventions in cattle production in colonial Niger. During the 1930s, the colonial authorities tried to transform nomadic pastoralists into sedentary farmers. A key element of their strategy was to introduce and breed cattle that could produce milk for urban markets. Krätli analyzes cultural contestations surrounding the "ideal" breed type, showing how WoDaaBe nomads, living in the precarious arid environment of the Sahel, strove to retain their Bororo cattle, which were adapted to withstand drought and seasonal shortages of grazing, thus illuminating scientific and popular practices in cattle breeding. As in many European colonies, the practice of veterinary medicine was as much about reordering indigenous society as it was about controlling disease.

Robert Perrins's chapter provides a welcome addition because it extends the scope of the collection beyond the Western world and the European colonies. His examination of the development of veterinary medicine by the Japanese in Manchuria introduces a new political and geographical dimension. In Manchuria, the development of veterinary services, as well as bacteriological institutions to investigate a number of local diseases, was viewed by the authorities as essential for Japanese settlement in northern China. The emphasis on creating and improving a settler economy, as opposed to prioritizing that of the indigenous people, mirrored similar episodes in some of the European colonies. Further extending the geographic scope of this volume, Rita Pemberton paints a more positive picture of the rise of state veterinary services in Trinidad and Tobago. She demonstrates how the threat of zoonoses was an important motivation for veterinary development. Nevertheless, British efforts to advance the livestock sector in Trinidad and Tobago were a response to the declining profits that European planters accrued from sugar production and were thus aimed at the ruling colonial elite.

In the European colonies, as well as countries in Europe and North America, the rise of the veterinary regime was not welcomed by all, and the same was true in the colonies of settlement. Recent studies on southern Africa have shown that the imposition of veterinary regulations was politically controversial, producing conflict between modernizing, "progressive" producers and subsistence farmers. Commercial agriculturists, as well as subsistence pastoralists who practiced transhumance to optimize grazing, often resented local quarantines and stock regulations if these meant that they could not transport their animals to market or move their livestock seasonally to desirable pastures.[31] The chapters by John Fisher and Robert Peden, set in late nineteenth-century Australasia, provide a useful counterpoint to the southern African case. Here the emphasis is on settler farmers, rather than veterinary practitioners and institutions. Fisher shows how wool producers in Australia, linked to metropolitan markets through the export trade, became increasingly concerned with scab in sheep. This condition arose from the gnawing of the acari mite and could result in considerable damage to the fleece. During the mid-nineteenth century, it was farmers, rather than the state, who experimented with dips and through their agricultural boards introduced local regulations that led to the eradication of the disease through regular insecticidal dipping. Fisher thus illustrates how veterinary science was part of a broader, progressive agenda set by colonial farmers, rather than necessarily being an imposition of officialdom.

In a chapter that provides thematic parallels, Robert Peden shows how New Zealand sheep farmers used selective breeding to eliminate a disease known as footrot. The standard wool-producing sheep, the Merino, was very susceptible under local conditions, and breeders responded by developing the Corriedale variety that was more tolerant of damp grazing lands. In contrast to Krätli's study of Niger, farmers rather than veterinarians took the lead in these breeding experiments. A comparison with South Africa, where progress along these lines took much longer, suggests that the possibilities for disease control were restricted not only by environmental contingency or limitations in scientific knowledge; local political, economic, social, and cultural factors have also played a role and have historically contributed to a variety of opportunities and outcomes in the management of livestock diseases.

Thus, overall, the historical presentations in this book focus primarily on the political economy of certain livestock diseases as well as on environmental issues pertaining to animal health. A subject that historians have been slower to respond to, however, is the epistemology of science itself. In fact, discussions about developments in veterinary science have largely remained a monopoly of practicing scientists, and only the laboratory revolution of the late nineteenth century, along with its political and social impacts, has engaged widespread attention from historians.[32] In general, the chapters here show how science was adopted by farmers and states as a tool of development, but little has been written about how the scientific knowledge that they used had been acquired or constructed. Yet the potential for developing this theme is considerable. The editors of this book have recently looked at the history of the Onderstepoort Veterinary Laboratories in South Africa, concentrating specifically on the type of science carried out at that institute, not just in the context of the political and economic agendas that underpinned veterinary research but also the actual work scientists themselves carried out in the laboratory and the field.[33] They have explored developments in microbiology and the discovery of vaccines, the ecology and control of arthropod borne diseases, and the dangers of plant poisonings, thereby giving scientists direct agency in the construction of veterinary knowledge. Similar studies are appearing for other institutions such as the Animal Research Station in Cambridge (U.K.).[34] The nature and evolution of veterinary science as a discipline, as well as further examinations of specific infections and ecologies of disease, in the format of either individual monographs or comparative studies, proffer exciting topics for further research by environmental and scientific historians alike.

Notes

1. For example, Iain Pattison, *The British Veterinary Profession, 1791–1948* (London: J. A. Allen, 1983); John Fisher, "Not Quite a Profession: The Aspirations of Veterinary Surgeons in England in the Mid-Nineteenth Century," *Historical Research* 66, no. 161 (1993): 284–302; Joanna Swabe, *Animals, Disease and Human Society: Human-Animal Relations and the Rise of Veterinary Medicine* (London: Routledge, 1999); Daniel Gilfoyle, "Veterinary Science and Public Policy at the Cape, 1877–1910" (DPhil thesis, University of Oxford, 2002); Susan Jones, *Valuing Animals: Veterinarians and Their Patients in Modern America* (Baltimore, MD: Johns Hopkins University Press, 2003).

2. Anne Hardy, "Professional Advantage and Public Health: British Veterinarians and State Veterinary Services, 1865–1939," *Twentieth Century History* 14, no. 1 (2003): 1–23; idem, "Pioneers in the Victorian Provinces: Veterinarians, Public Health and Urban Animal Economy," *Urban History* 29, no. 3 (2002): 372–87; Keir Waddington, "'Unfit for Human Consumption': Tuberculosis and the Problem of Infected Meat in Late Victorian Britain," *Bulletin of the History of Medicine* 77, no. 3 (2003): 636–61; idem, "To Stamp Out 'So Terrible a Malady': Bovine Tuberculosis and Tuberculin Testing in Britain 1890–1939," *Medical History* 48, no. 1 (2004): 29–48.

3. John Fisher, "Cattle Plagues Past and Present: The Mystery of Mad Cow Disease," *Journal of Contemporary History* 33, no. 2 (1998): 215–28; Abigail Woods, "The Construction of an Animal Plague: Foot and Mouth Disease in Nineteenth-Century Britain," *Social History of Medicine* 17, no. 1 (2004): 23–39; idem, "'Flames and Fear on the Farms': Controlling Foot and Mouth Disease in Britain, 1892–2001," *Historical Research* 77, no. 198 (2004): 520–42; idem, *A Manufactured Plague: The History of Foot and Mouth Disease in Britain* (London: Earthscan, 2004).

4. John Fisher "A Pandemic (Panzootic) of Pleuropneumonia, 1840–1860," *Historia Medicinae Veterinariae* 11, no. 1 (1986): 26–32; idem, "To Kill or Not to Kill: The Eradication of Contagious Bovine Pleuro-pneumonia in Western Europe," *Medical History* 47, no. 3 (2003): 314–31. On rinderpest, see, for example, Michael Worboys, "Germ Theories and British Veterinary Medicine, 1860–1890," *Medical History* 35, no. 3 (1991): 308–27; idem, "Veterinary Medicine, the Cattle Plague and Contagion, 1865–1890," in *Spreading Germs: Disease Theories and Medical Practice in Britain, 1865–1900* (Cambridge: Cambridge University Press, 2000), 43–72; C. Huygelen, "The Immunization of Cattle against Rinderpest in Eighteenth-Century Europe," *Medical History* 41, no. 2 (1997): 182–96. For rinderpest in South Africa, see Charles van Onselen, "Reactions to Rinderpest in Southern Africa, 1896–97," *Journal of African History* 13, no. 3 (1972): 473–88; Pule Phoofolo, "Epidemics and Revolutions: The Rinderpest Epizootic in Late Nineteenth-Century Southern Africa," *Past and Present* no. 138 (February 1993): 112–43; and Daniel Gilfoyle, "Veterinary Research and the African Rinderpest Epizootic: The Cape Colony, 1896–98," *Journal of Southern African Studies* 29, no. 1 (2003): 133–54.

5. Andrew Cunningham and Perry Williams, eds., *The Laboratory Revolution in Medicine* (Oxford: Oxford University Press, 1992); Worboys, *Spreading Germs*.

6. Karen Brown, "Poisonous Plants, Pastoral Knowledge and Perceptions of Environmental Change in South Africa, c. 1880–1940," *Environment and History* 13, no. 3 (2007): 307–32.

7. For example, John Ford, *The Role of Trypanosomiasis in African Ecology: A Study of the Tsetse Fly Problem* (Oxford: Clarendon, 1971); Leroy Vail, "Ecology and History: The Example of Zambia," *Journal of Southern African Studies* 3, no. 2 (1977): 129–55; John McCracken, "Experts and Expertise in Colonial Malawi," *African Affairs* 81, no. 322 (1982): 101–16; idem, "Experts and Amateurs: Tsetse, Nagana and Sleeping Sickness in East and Central Africa," in *Imperialism and the Natural World*, ed. John MacKenzie (Manchester: Manchester University Press, 1990), 187–212; James Giblin, "Trypanosomiasis Control in African History: An Evaded Issue?," *Journal of African History* 31, no. 1 (1990): 59–80; Richard Waller, "Tsetse Fly in Western Narok, Kenya," *Journal of African History* 31, no. 1 (1990): 81–101; Helge Kjekshus, *Ecology Control and Economic Development in East African History: The Case of Tanganyika, 1850–1950* (London: James Currey, 1996); Kirk Arden Hoppe, *Lords of the Fly: Sleeping Sickness Control in British East Africa, 1900–1960* (Westport, CT: Praeger, 2002); Karen Brown, "From Ubombo to Mkhuzi: Disease, Colonial Science and the Control of Nagana (Livestock Trypanosomosis) in Zululand, South Africa, c. 1894–1953," *Journal of the History of Medicine and Allied Sciences* 63, no. 3 (2008): 285–322. For an alternative and broader environmentalist approach, see William Beinart, "Vets, Viruses and Environmentalism at the Cape," *Paideuma* 43. (1997): 227–52.

8. Swabe, *Animals, Disease and Human Society*, 11.

9. Ibid.; Jones, *Valuing Animals*; Karen Brown, "Tropical Medicine and Animal Diseases: Onderstepoort and the Development of Veterinary Science in South Africa, 1908–1950," *Journal of Southern African Studies* 31, no. 3 (2005): 413–529.

10. Alan Marcus, *Agricultural Science and the Quest for Legitimacy: Farmers, Agricultural Colleges and Experiment Stations, 1870–90* (Ames: Iowa State University Press, 1985).

11. J. F. Smithcors, *The American Veterinary Profession: Its Background and Development* (Ames: Iowa State University Press, 1963), chap. 12; C. K. Hutson, "Texas Fever in Kansas, 1866–1930," *Agricultural History* 68, no. 1 (1994): 74–104; Susan Jones, "Laboratory Science and Common Sense," *Veterinary Heritage* 22, no. 2 (1999): 25–30.

12. Paul F. Cranefield, *Science and Empire: East Coast Fever in Rhodesia and the Transvaal* (Cambridge: Cambridge University Press, 1991); Gilfoyle, "Veterinary Science and Public Policy at the Cape, 1877–1910"; Brown, "From Ubombo to Mkhuzi."

13. For U.S. progressivism and scientific development, see, for example, Samuel P. Hays, *The Response to Industrialism, 1885–1914* (Chicago: University of Chicago Press, 1957); idem, *Conservation and the Gospel of Efficiency: The Progressive Conservation Movement, 1890–1920* (Cambridge, MA: Harvard University Press, 1959); Charles Rosenberg, *No Other Gods: On Science and American Social Thought* (Baltimore, MD: Johns Hopkins University Press, 1976); idem, "Rationalization and Reality in the Shaping of American Agricultural Research, 1875–1914," *Social Studies of Science* 7, no. 4 (1977): 401–22; Elizabeth Sanders, *Roots of Reform: Farmers, Workers and the American State, 1877–1917* (Chicago: University of Chicago Press, 1999). On British ideas about "national efficiency," see Bernard Semmel, *Imperialism and Social Reform: English Social Imperial Thought, 1895–1914* (London: G. Allen and Unwin, 1960); Michael Worboys, "Science and British Colonial Imperialism, 1895–1940" (PhD. thesis, University of Sussex, 1979); John Clark, "Science, Secularization and Social Change: The Metamorphosis of Entomology in Nineteenth-Century England" (DPhil thesis, University of Oxford, 1994).

14. Gerald L. Geison, *The Private Science of Louis Pasteur* (Princeton, NJ: Princeton University Press, 1995); Thomas D. Brock, *Robert Koch: A Life in Medicine and Bacteriology* (Madison, WI: Science Tech Publishers, 1988).

15. Worboys, *Spreading Germs*, 43–72.

16. Fisher, "A Pandemic (Panzootic) of Pleuropneumonia, 1840–1860"; Jeffrey B. Peires, *The Dead Will Arise: Nongqawuse and the Great Xhosa Cattle Killing Movement of 1856–57* (Johannesburg: Ravan, 1989); Christian Andreas, "The Lungsickness Epizootic of 1853–1857: An Analysis of Its Socio-Economic Im-

pact and the Ensuing Reactions in the Cape Colony and Xhosaland" (master's thesis, University of Hannover, 2003); Fisher, "To Kill or Not to Kill."

17. James Giblin, "East Coast Fever in Socio-historic Context: A Case Study from Tanzania," *International Journal of African Historical Studies* 23, no. 3 (1990): 401–21; Cranefield, *Science and Empire.*

18. Alfred Crosby, *The Columbian Exchange: Biological and Cultural Consequences of 1492* (Westport, CT: Greenwood, 1972); idem, *Ecological Imperialism: The Biological Expansion of Europe, 900–1900* (Cambridge: Cambridge University Press, 1986); see also William McNeill, *Plagues and Peoples* (Oxford: Blackwell, 1977).

19. Van Onselen, "Reactions to Rinderpest"; Phoofolo, "Epidemics and Revolutions"; Gilfoyle, "Veterinary Research and the African Rinderpest Epizootic."

20. Clive A. Spinage, *Cattle Plague: A History* (New York: Kluwer Academia, 2003).

21. On the nature of tropical medicine and its institutional form, see P. Manson-Bahr, *Patrick Manson: The Father of Tropical Medicine* (London: Thomas Nelson and Sons, 1962); Michael Worboys, "Manson, Ross and Colonial Medical Policy: Tropical Medicine in London and Liverpool, 1819–1914," in *Disease Medicine and Empire: Perspectives on Western Medicine and the Experience of European Expansion,* ed. Roy MacLeod and Milton Lewis (London: Routledge, 1988), 21–37; idem, "Germs, Malaria and the Invention of Mansonian Tropical Medicine: From 'Diseases in the Tropics' to 'Tropical Diseases,'" in *Warm Climates and Western Medicine,* ed. David Arnold (Amsterdam: Rodopi, 1996), 181–207; D. Haynes, *Imperial Medicine: Patrick Manson and the Conquest of Tropical Disease* (Philadelphia: University of Pennsylvania Press, 2001).

22. Brown, "From Ubombo to Mkhuzi," 293–98.

23. Diana K. Davis, "Prescribing Progress: French Veterinary Medicine in the Service of Empire," *Veterinary Heritage* 29, no. 1 (2006): 3.

24. For example, see Cranefield, *Science and Empire;* Brown, "Tropical Medicine and Animal Diseases."

25. Most notably on trypanosomosis, Ford, *The Role of Trypanosomiasis in African Ecology;* Giblin, "Trypanosomiasis Control in African History"; Kjekshus, *Ecology Control and Economic Development in East African History.*

26. For an overview, see John M. Mackenzie, *Empire of Nature: Hunting, Conservation and British Imperialism* (Manchester: Manchester University Press, 1988).

27. Davis, "Prescribing Progress," 5.

28. See, for example, Donald Worster, *Dust Bowl and the Southern Plains in the 1930s* (Oxford: Oxford University Press, 1979); William Beinart, "Soil Erosion, Conservation and Ideas about Development: A Southern African Exploration, 1900–1960," *Journal of Southern African Studies* 11, no. 1 (1984): 52–83;

Ian Phimister, "Discourse and Discipline of Historical Context: Conservationism and Ideas about Development in Southern Rhodesia, 1930–1950," *Journal of Southern African Studies* 12, no. 2 (1986): 263–75; David Anderson, *Eroding the Commons: The Politics of Ecology in Baringo, Kenya, 1890–1963* (Oxford: James Currey, 2002); Belinda Dodson, "Above Politics? Soil Conservation in 1940s South Africa," *South African Historical Journal* 50 (2004): 49–64.

29. Daniel R. Headrick, *The Tools of Empire: Technology and European Imperialism in the Nineteenth Century* (Oxford: Oxford University Press, 1981); idem, *The Tentacles of Progress: Technology Transfer in the Age of Imperialism, 1850–1940* (Oxford: Oxford University Press, 1988).

30. Important texts are David Arnold, ed., *Imperial Medicine and Indigenous Societies* (Manchester: Manchester University Press, 1988); Roy MacLeod and Milton Lewis, eds., *Disease, Medicine and Empire: Perspectives on Western Medicine and the Experience of European Expansion* (London: Routledge, 1988); Megan Vaughan, *Curing Their Ills: Colonial Power and African Illness* (Stanford, CA: Stanford University Press, 1991); David Arnold, *Colonizing the Body: State Medicine and Epidemic Disease in Nineteenth-Century India* (Berkeley: University of California Press, 1993); Mark Harrison, *Public Health in British India: Anglo-Indian Preventive Medicine, 1859–1914* (Cambridge: Cambridge University Press, 1994); and Andrew Cunningham and Bridie Andrews, eds., *Western Medicine as Contested Knowledge* (Manchester: Manchester University Press, 1997).

31. For South African examples, see Mordechai Tamarkin, "Flock and Volk: Ecology, Culture, Identity and Politics among Cape Afrikaner Stock Farmers in the Late Nineteenth Century" (paper presented at the conference "African Environments, Past and Present," Oxford, July 1999); William Beinart, *The Rise of Conservation in South Africa: Settlers, Livestock and the Environment, 1770–1950* (Oxford: Oxford University Press, 2003).

32. There are many works dealing with laboratory science associated with Louis Pasteur, Robert Koch, and other important scientists at their respective institutes, such as those by Elie Metchnikoff and Paul Ehrlich. Major works dealing with laboratory experiments and practices for this period, to name but two, are Geison, *Private Science of Louis Pasteur,* and Brock, *Robert Koch.*

33. For an overview, see Brown, "Tropical Medicine and Animal Diseases."

34. For example, Chris Polge, "The Animal Research Station in Cambridge" (paper presented at a conference titled "Between the Farm and the Clinic: Agriculture and Reproductive Technology in the Twentieth Century," University of Cambridge, 29 April 2005).

CHAPTER 1

Epizootic Diseases in the Netherlands, 1713–2002

Veterinary Science, Agricultural Policy, and Public Response

Peter A. Koolmees

THE GROWTH of livestock production has been regularly threatened and hampered by outbreaks of epizootic diseases, not only today but also in the past. The spread of contagious livestock diseases often coincided with animal movements due to trade or wars.[1] The path followed by the disease can be closely observed on a local scale, too. Because of the socioeconomic implications of livestock diseases on the human food supply, many archives document the measures taken by local and national authorities to prevent further spread of the disease and to deal with the economic consequences for the farmers.[2]

Over the last three centuries, the threat of epizootic diseases has grown due to the expansion of national and international livestock trade. This led to calls for effective prevention and control of livestock diseases. Several European countries accepted the challenge, establishing state veterinary services and cattle-disease control acts around 1900. Mass outbreaks of animal diseases also led to scientific developments, as well as to major state interventions in rural society.[3]

The outbreak of classical swine fever in the Netherlands in 1997 and particularly the epizootic of foot-and-mouth disease (FMD) in 2001–2 led to great societal commotion and criticism on intensive livestock farming. The public became outraged when it was regularly confronted with images of mass slaughter, not only of diseased livestock but also of sound animals, particularly since these diseases posed no threat to human health. Livestock producers, the European Union (EU), the Ministry of Agriculture, and veterinarians alike were subject to this criticism. The latter were forced to follow a strict cull-and-slaughter policy after the EU adopted a nonvaccination policy in 1991.

Despite superficial similarities, the response to recent outbreaks of epizootic diseases differed dramatically from previous responses. In this chapter, a comparison is made between major outbreaks of contagious livestock diseases that struck the Netherlands during the last three centuries. Attention will be paid to the role of veterinary science, the agricultural policy applied, and the public response. The choice and rationalization for particular control strategies will be discussed, as will potential reasons for the changing reaction to outbreaks of epizootic diseases over time. Before turning to the major outbreaks of epizootic diseases that struck the Netherlands, some general remarks will be made on the subject of animal plagues in history.

History of Epizootic Diseases

Many written sources are available with respect to information on the history of epizootic diseases.[4] This subject was even a separate discipline taught at veterinary schools throughout the eighteenth and nineteenth centuries. Most of these authors agree that, at least in European history, rinderpest[5] was always thought to have come from the East where it was endemic on the Russian steppes. The spread of contagious animal diseases such as rinderpest, contagious bovine pleuropneumonia (CBPP), and FMD was caused by international cattle trade and by armies that transported infected animals as victuals. For instance, the outbreak of rinderpest in the 1740s resulted from the Austrian War of Succession when infection passed from Hungary to other European countries. This traditional history of rinderpest, with its emphasis on the relationship between infection, warfare, and cattle trade, was criticized by J. A. Faber. According to him, outbreaks did not always coincide with war or spread along transport routes, and it is often difficult to diagnose rinderpest or any other epizootic after the fact. Further, we cannot be certain that the outbreaks described in historical documents are analogous to the devastating outbreaks we know from contemporary times.[6] On the other hand, based on present-day experience and knowledge of epidemiology, it is obvious that the international cattle trade and cattle drives must have contributed to the spread of contagious animal diseases.

From the numerous historical decrees concerning the control of rinderpest, it is clear that authorities were well aware of the potential threat to local livestock and cattle markets posed by cattle drives. National, regional, and local authorities alike issued special rinderpest decrees to prevent local livestock from being exposed to outside sources of infection. These decrees were particularly aimed at avoiding contact between local herds and caravans of imported cattle (including their owners and drovers). The overland oxen routes were often located away from the main (state) roads. Also, contact with the local human and animal populations was avoided as much as possible at the inns along the transport routes and during the purchase of fodder from local farmers. In the second half of the eighteenth century, a special veterinary police service was established in Austria. As part of the cordon sanitaire, these mounted civil servants patrolled the long eastern border to prevent illegal cattle imports; the veterinary police also supervised quarantine measures. These measures, though, were inadequate, and rinderpest outbreaks still occurred regularly. In the beginning of the nineteenth century, a German veterinarian described how asymptomatic cattle originating in Russia and the Ukraine still infected the highly sensitive livestock populations in Poland and East Prussia.[7]

Apart from the veterinary policy measures mentioned above, some interesting questions with respect to the oxen trade and animal diseases still remain. From a historical point of view, it would be interesting to determine if herdsmen and drovers provided medical treatments or therapies to the animals in their care. If so, were these treatments based on experience, on popular veterinary medicine, or on contemporary medical or veterinary sources? The same question could be asked of the decrees and measures issued by governments: were these based on expert advice or on experience? A systematic study of available historical resources on animal trade and transport could address some of these questions.[8]

Before dealing with outbreaks of epizootics in the Netherlands, some background on livestock production in that country over time will be presented in order to give a better understanding of the impact that such diseases had on society. Livestock production was traditionally a large contributor to the Dutch economy. A mild sea climate, plenty of fertile grasslands, extended rail-, road- and waterway systems and seaports presented favorable conditions for livestock production and exports of foods of animal origin, particularly with large markets in London, Paris, and Hamburg at close distance. Therefore, from the late Middle Ages onward, the Netherlands specialized in livestock production and, especially, exports of meat, butter, and cheese. Due to the opening of the British market to foreign imports in

1842, London became the main market for Dutch meat and dairy exports until 1940. Cattle and sheep represented the primary livestock species until 1870; from then on, pork production increased rapidly. Sheep have always been produced largely for export purposes. Dairy-cows have always constituted the vast majority of the cattle population. Between 1800 and 1990, numbers of cattle and pigs increased from 1.1 to 4.7 million and from 0.3 to 14 million, respectively. Due to environmental regulations on nitrogen input, fixed milk quotas, and public criticism of industrialized farming and its negative consequences for animal welfare and the environment, livestock numbers began to decrease in the late 1980s and continue to do so.[9]

Now it is time to have a closer look at a cattle disease that was most feared over the centuries.

Rinderpest in the Netherlands

Rinderpest (*Pestis bovina*) in ruminants is caused by a morbillivirus, an organism closely related to the causative agents of measles and canine distemper. Transmission requires direct or close indirect contact; infection is via the nasopharynx. The incubation period is three to fifteen days. Virulence varies between strains but, during epizootics, the morbidity rate is often 100 percent, and the mortality rate ranges between 60 and 90 percent. It is the most lethal plague known in cattle. Animals that survive infection develop a high level of long-lasting immunity. Clive Spinage states that virus carriage is a very temporary state and that the development of a persistent carrier of the virus is very rare. Nevertheless, he suggested that, in the eighteenth century, rinderpest could have been introduced into South Africa from the Netherlands—where by that time it was endemic—via virus carriers after a four-month journey by ship.[10]

Between 1713 and 1867, the Netherlands was struck four times by a major outbreak of rinderpest. These outbreaks affected large areas, but smaller local outbreaks also occurred over this time period (see table 1.1). The consequences of such outbreaks were deeply felt in Dutch society as a whole, and rinderpest was greatly feared. Confronted with outbreaks of animal distemper, religious authorities organized days of public prayers. Local governments issued various decrees, including an embargo on cattle imports, a ban on livestock movement from infected areas and cattle markets, and detailed instructions for the disposal of animals that had died from the disease. These measures probably prevented the spread of epizootics to some extent. The same can be said for inoculation, which was applied in the second half of the eighteenth century. However, the only effective remedy to an outbreak was the immediate slaughter of all infected animals

Table 1.1. Number of dead and slaughtered bovines
during outbreaks of rinderpest in the Netherlands

Period	Area	Slaughtered bovines	Mortality
1713–20	Whole country	None	120,000
1744–65	Whole country	None	1,000,000
1768–86	Whole country	None	800,000
1813–14	Utrecht	300	50
1865–67	Holland and Utrecht	27,000	78,000

Sources: Jan Bieleman, Geschiedenis van de landbouw in Nederland, 1500–1950 (Meppel, the Netherlands: Boom, 1992), 166, 291; Piet D. 't Hart, "Pestis bovina in Utrecht, 1813–1814," Maandblad Oud-Utrecht 46, no. 1 (1973): 4; Cees Offringa, Van Gildestein naar Uithof: 150 jaar diergeneeskunde onderwijs in Utrecht (Utrecht: Faculteit der Diergeneeskunde, 1971), 113.

and animals that were suspected of carrying the disease. This well-known cull-and-slaughter or Lancisi system was introduced by Giovanni Maria Lancisi in 1711 in Italy and by Thomas Bates in 1714 in England. Before that time, the authorities had to rely on veterinary policy measures of isolation, containment, and quarantine of infected animals.[11]

In June 1713, the initial victims of the first major eighteenth-century outbreak of rinderpest were reported around Amsterdam (province of Holland). Slaughter oxen imported from Denmark probably caused the outbreak. In spite of an import embargo and a containment order, the disease spread to the provinces of Utrecht and Friesland. In 1715, cases occurred all over the country. Local and provincial authorities instituted no control measures except embargos on imports, transportation, and markets. The disease remained in the country until 1720. The epizootic of 1744 was the most severe one, causing the death of about one million animals in the course of twenty years. Again, the disease started in the province of Holland and spread from there to most other regions of the country. The measures taken by the authorities did not differ from those taken in the 1713 epizootic. In 1768, rinderpest started in the northeastern part of the country, but eventually the whole country was infected again. In total, mortality amounted to eight hundred thousand bovines.[12]

During the eighteenth century, about two million cattle died of rinderpest over a period of fifty years. The pattern of each epizootic was the same. After introduction, the disease spread rapidly and killed most animals in the first two years. Then mortality quickly dropped because fewer animals remained and a majority of those had obtained immunity. The disease prevailed in the high-density cattle areas of the provinces of Holland, Utrecht, and Friesland. The chance of rinderpest reintroduction remained low as

Figure 1.1. Lithograph by Jan Smit, 1745: "The hand of God struck the Netherlands, afflicting its cattle with rinderpest." Amsterdam, S. Van Esveldt and J. Maagh Alkmaar, booksellers, 1746. Courtesy of University of Utrecht Library

long as trade barriers were maintained by official measures. Imposition of official measures, though, was hampered by the lack of a strong central government in the eighteenth-century Dutch Republic. Local or regional authorities imposed trade barriers to contain the disease but did so according to local interests and with no view as to preventing disease across the country. Religious and political-ideological objections were brought forward against local measures that were considered too rigorous. As soon as the ban on cattle imports was lifted, rinderpest occurred again. The repetition of this pattern explains why the disease remained an issue for so long.[13]

As would be expected from this rinderpest example, animal plagues had a significant impact on the rural economy, on society as a whole, and on veterinary medicine.

Rural Economy and Public Response

Conventional wisdom holds that animal diseases forced many farmers to quit their business—a view not in keeping with recent research that shows a majority of farmers surviving the crisis. The Dutch rural economy was more dynamic than often claimed in the nineteenth- and early twentieth-century agricultural historiography. Although the misery of the farmers initially seemed to be insurmountable, most of them were able to continue their farming by becoming inventive entrepreneurs. They instituted, for instance, a temporary shift to cheese production, sheep breeding, fattening of calves, or arable farming. Many farmers profited from higher meat and dairy prices during epizootics. All this resulted in a surprisingly fast recovery of the cattle stock and an over-all improvement in the farm economy. By the end of the eighteenth century, farmers owned considerably more land than had been owned around 1700.[14]

Until well into the eighteenth century, outbreaks of contagious livestock diseases were considered a divine punishment for a sinful population. The general opinion was that one simply had to endure one's fate. During outbreaks, prayer meetings were held, and the end of the outbreak was celebrated with a general thanksgiving day in all churches. Superstition also played a role. In Denmark, for instance, a calf from each herd was buried alive, or cattle were driven through a fire to prevent infection. In addition, all kinds of folk remedies were applied, especially the administration of various kinds of potions and herbs.

Scientific Progress

The rinderpest epizootics stimulated the development of veterinary science and the application of treatments ranging from hygiene and quarantine measures, polypharmacy, to cull and slaughter, inoculation, and vaccination. For instance, during the 1744 epizootic, the States of Holland approached the medical faculty of Leiden University for advice. In 1745, the municipality of the city of Utrecht asked medical professors at the local university to investigate the disease and provide advice on how to prevent or cure it. Before the emergence of veterinary schools in the late eighteenth century, though, this approach was more an exception than a rule. By then, the Enlightenment stimulated a more scientific approach to societal problems, including epizootics. Initiatives were typically driven by individuals rather than by universities with Cornelius Nozeman, a clergymen working in the countryside, and Geert Reinders, a learned farmer from Groningen, being but two of many examples.[15] A few persons with a scientific background also performed experiments with inoculations, such as the physicians Pieter Vink from Rotterdam and Petrus Camper from Groningen.[16] The publications of the latter were well known at home as well as abroad. Overall, the inoculation experiments were not very successful, suffering from a lack of uniformity in method and hygiene.[17] Abroad, the systematic killing of all infected and suspected animals (cull and slaughter) was considered a more effective method. In countries with a strong central government where a cull-and-slaughter policy was rigorously executed, infectious livestock diseases were controlled with minimal losses. This result was even more impressive when one considers that the germ theory had not yet been developed.[18] Overall, the rinderpest epizootics contributed significantly to the establishment of the first veterinary schools in the second half of the eighteenth century.

Rinderpest in the Nineteenth Century

Under the Napoleonic occupation, the Netherlands had become a unified state—an approach also taken with measures issued against epizootics. As a result of the last eighteenth-century outbreak of rinderpest, a minor one in the provinces of Gelderland and Zeeland (1796–99), the first national Livestock Act was enacted in 1799. This act prescribed both the obligation to notify the authorities in case of a contagious cattle disease and the cull and slaughter of infected cattle. Inoculation was prohibited. In the same year, the Cattle Fund was established to indemnify farmers for losses among their herds. The new rules were applied successfully in 1813 in Utrecht when rinderpest was introduced by Prussian troops. The police and military closed the area around the infected farm with a cordon sanitaire, while all diseased and suspected animals were killed. In June 1814, all bans on transports and markets were lifted; two weeks later, the newspapers reported that no new cases had occurred.[19]

The severe outbreak of rinderpest from 1865 to 1867 changed the government's attitude toward eradication strategies and veterinary medicine. In July 1865, rinderpest was introduced again in Holland by the reimportation of Russian oxen for which no customer could be found in England. Provincial and municipal councils failed to take adequate measures, and the disease started spreading on a large scale. Cattle exports, which represented considerable earnings for the agricultural sector, decreased dramatically when foreign countries closed their borders. Again, the government hesitated to take drastic and decisive measures. This failure was caused, in part, by contradictory advice provided by veterinary committees. The credibility of veterinary science as well as that of the Dutch veterinary profession, was questioned in both the parliament and the newspapers.[20]

It took more than a year of disagreement between various committees and a change from a liberal to a conservative government before mandatory slaughter of all diseased and suspected animals was begun. Even then, many difficulties were encountered. The farmers simply could not be persuaded to submit to the harsh regulations regarding expropriation, slaughter, burial of dead animals, and cleaning of stables. They had to be forced to do so by the infantry, the cavalry, the navy, the artillery, and the law. Still, a few offenders were shot dead while smuggling animals at night or openly resisting the enforcement of police measures because they did not want to

violate the will of God.[21] Nevertheless, a rigorous cull-and-laughter policy was carried out; farmers were fully compensated for their sound and
suspect animals and at 60 percent of the value for diseased cattle. Within
three months, the disease disappeared from the Netherlands despite the fact
that, in the infected areas, a quarter of all bovines were infected. Between
July 1866 and December 1867, seventy-eight thousand bovines died, while
more than twenty-seven thousand were killed. The total economic damage
of the outbreak amounted to thirteen million Dutch guilders. The merits
of veterinary committees, which repeatedly advised carrying out a drastic
cull-and-slaughter strategy, were acknowledged. The successful growth of
the veterinary service culminated in the enactment of the Livestock Act in
1870 and the establishment of a national Veterinary Service, although that
measure was adopted in the Dutch Parliament by a vote of only thirty-
two to thirty. This act was based on the recommendations of an international veterinary congress held in Zürich in 1867.[22] Since 1867, rinderpest
has been considered eradicated from the Netherlands. Attempts to combat
outbreaks of other epizootic diseases in the nineteenth century, however,
took more effort.

Contagious Bovine Pleuropneumonia

Contagious bovine pleuropneumonia (*Pleuropneumonia contagiosa*), a highly
contagious disease caused by *Mycoplasma mycoides*, presented a major
problem for both farmers and veterinarians. Often mistaken for anthrax
or rinderpest, CBPP is a mostly subacute or chronic affection with an incubation time of three to eight weeks. Susceptible cattle become infected by
inhaling droplets disseminated by coughing in affected cattle. As would be
expected with such an incubation period, epizootics follow a slow course.
Morbidity rates of 10 percent are most common; the mortality rate is about
50 percent. Of recovered animals, 25 percent may become carriers.[23]

In 1831, the disease was first diagnosed in the Netherlands. Through
1887, about 250,000 bovines died of the disease. This outbreak led to an
increased call for veterinary state supervision in the 1840s and 1850s. Exhaustion of the Cattle Fund in 1849 by payments made to the farmers with
infected herds also played a role. However, there was a strong disagreement concerning the strategy to combat CBPP between contagionists and
anticontagionists. The veterinarian Jacob van Hertum, who worked in
Zeeland, a province then existing of several islands, successfully applied
a system of cull and slaughter and containment. However, Alexander Nu-
man, the director of the State Veterinary School in Utrecht, disagreed with
this approach, stating that the contagious nature of CBPP was still unclear.

Consequently, veterinarians were not able to provide the government with firm advice on its policy.[24] In addition, there was strong opposition against a veterinary service by physicians; they questioned the scientific level of veterinary medicine. Why should acts for veterinary state supervision with cattle be established while a system of coherent medical acts was still lacking? Furthermore, the liberal constitution adopted in 1848 placed great emphasis on local autonomy, further hampering the creation of a national veterinary service. Municipal and provincial boards were held primarily responsible when calamities such as livestock diseases occurred.

Similar to physicians who were powerless against cholera, veterinarians initially had no cure for CBPP. Physicians, though, were considered academic men, while veterinarians were judged only on their economic merit and ability to keep livestock healthy. In 1852, the Belgian physician Louis Willems published the positive outcome of his inoculation experiments in cattle infected with CBPP. His method was based on the inoculation of infectious matter obtained from the lungs of infected animals into the tails of sound cattle. After experiments with the Willems method by the State Veterinary School, the Dutch government supported inoculations in infected areas. Based on the Cattle Act of 1870, a campaign against CBPP combining slaughtering and inoculation was started in 1878. The disease was last diagnosed in the Netherlands in 1887. Between 1831 and 1887, the total costs of eradication amounted to six million Dutch guilders.[25]

Compared to rinderpest, CBPP did not cause great societal commotion; it was seen more as a problem for the rural economy and as an opportunity for scientific debate. This may be due to the chronic nature of the disease and a lower mortality rate than rinderpest. The dispute over whether meat originating from animals infected with CBPP could be consumed safely was less fierce than the debate over the safety of meat from animals with rinderpest. Initially, much meat from animals infected with rinderpest was buried. Meat originating from bovines that had died of CBPP, though, found its way more easily to consumers.[26]

Similar to CBPP, classical swine fever did not cause great societal commotion initially. This changed in the course of the twentieth century, when outbreaks of swine fever led to the killing of huge numbers of pigs.

Classical Swine Fever

Swine fever represented another major epizootic disease that farmers and veterinarians had to deal with. Classical swine fever (*Pestivirus flaviviridae*) is a viral infection, although for a long time American researchers claimed it was caused by *Bacillus cholerae-suis*. The disease has acute and chronic forms, and virulence varies from severe, with high mortality, to mild or even subclinical. The incubation period is typically two to six days, with death at ten to twenty days after infection. The main source of infection is the pig, either live animals or uncooked pig products. In its acute form, the disease generally results in high morbidity and mortality.[27] The veterinary bacteriologist Jan Poels first diagnosed swine fever in the Netherlands in 1899. His pioneering research led to the foundation of the National Serum Institute in 1904. This institute worked closely together with the Veterinary Service. Many experiments were performed with serums against classical swine fever and FMD. Around 1900, the production of effective vaccines against these diseases was facilitated by microbiological advances.[28] However, this did not mean that cull and slaughter disappeared as an important eradication strategy.

A systematic eradication plan with compulsory notification was begun only in 1936, when swine fever was added to the Livestock Act. Meanwhile, research aimed at developing an effective vaccine continued. In 1961, a major outbreak was combated with a combination of vaccination, isolation, and cull and slaughter. During that outbreak, more than 320,000 animals were killed (table 1.2). From 1967 onward, only cull and slaughter was used.[29]

In February 1997, the Netherlands was confronted with a severe epidemic of classical swine fever. In September of that year, four hundred farms were infected. Despite the availability of an effective vaccine against swine fever, the nonvaccination policy of the EU dictated that Dutch authorities rely on a cull-and-slaughter policy. The EU policy was based on the fact that meat from vaccinated hogs was seropositive for swine fever and veterinarians were unable to determine whether the seropositivity was due to natural infection or vaccination. Major importing countries like Japan and the United States had swine-fever-free markets and refused entry of pork testing seropositive for swine fever. On 22 March 1997, a total ban on exports of living pigs was issued. About 650,000 pigs from infected farms were killed; more than one million were killed preventively. As a result of the transport ban, almost eight million healthy piglets had to be killed due to overpopulation in stables. In total, more than 9.6 million pigs lost their lives during this crisis. This was the biggest outbreak of swine fever in the Netherlands ever and the biggest swine-fever outbreak ever in the EU as a whole.[30]

Apart from the economic damage, which was estimated at five billion Dutch guilders, the epidemic resulted in broad public criticism of intensive pig farming and of veterinarians who had to kill healthy animals. It also led to tensions and concern within the veterinary profession, particularly between swine practitioners and official veterinarians responsible for the execution of the cull-and-slaughter policy. The swine-fever epidemic negatively influenced the image of animal production in general. Discussions on animal welfare and sustainable animal production brought about a change in the policy concerning livestock production. The Dutch government issued several measures to further reduce livestock numbers.[31] However, the poor image of intensive animal production was to be even more negatively influenced by severe outbreaks of another feared scourge, namely, foot-and-mouth disease.

Foot-and-Mouth Disease

Foot-and-mouth disease (*Aphthae epizooticae*) is a highly infectious viral disease of cattle, pigs, sheep, and buffalo. FMD is caused by an apthovirus of the family Picornaviridae involving seven distinct serotypes. Transmission of FMD is generally by contact between susceptible and infected animals. The incubation period is from two to fourteen days. Although morbidity approaches 100 percent, the lethality rate is typically only 2 to 5 percent. Nevertheless, FMD is considered an important livestock disease because of its economic impact (loss of productivity and decrease in animal weight).[32] FMD has been one of the listed epizootic diseases in the Dutch Livestock Act from 1880 onward. The major outbreaks in 1911, 1924, and 1938 affected many farms, prompting the government to pay many reimbursements. During the outbreak of 1911, for instance, livestock in more than seventy-one thousand farms were infected; eleven thousand bovines, three hundred goats and sheep, and sixty-five hundred pigs died of the disease, while almost five thousand animals were killed preventively. Between 1880 and 1925, a mixed policy of cull and slaughter and just leaving the disease run its course was followed. Significant progress in dealing with FMD was made after the establishment of the State Veterinary Research Institute in Rotterdam in 1929. There, Dr. Herman Salomon Frenkel and his coworkers developed and produced an effective vaccine against FMD in 1935.[33]

In the post–World War II decades, campaigns against FMD were quite successful. The campaigns provided a lot of diagnostic and preventive work for practitioners. Provincial animal-health services proved to be very efficient and successful in organizing campaigns against FMD and swine fever. An outbreak of FMD in 1953 was successfully eradicated by a new vaccine, again developed by Frenkel and his colleague M. van Waveren. An annual preventive vaccination for all bovines was started in 1954. In 1959, the National Serum Institute and the State Veterinary Research Institute were united in the Central Veterinary Institute. One of the main tasks of this institute remained the development of therapies against epizootic diseases. From 1960 onward, a change in cull-and-slaughter policy occurred.

Table 1.2. Number of infected farms with FMD and classical swine fever and number of animals killed during major outbreaks in the Netherlands

Period	Disease	Infected farms	Animals killed
1911	FMD	71,325	5,000
1924	FMD	88,930	5,000
1938	FMD	112,886	11,000
1961–62	FMD	5,647	322,000
1997–98	Swine fever	429	9,600,000
2001–2	FMD	26	280,000

Sources: K. G. Robijns, "Swine Fever and Swine Fever Eradication in the Netherlands," in *Veterinary Work in the Netherlands* (The Hague: Ministry of Agriculture, 1970), 180–85; *Verslagen Veeartsenijkundig Staatstoezicht* over 1911, 118; 1924, 70; 1938, 150; 1961, 58; Dick J. Vervoorn, "Control of Foot-and-Mouth Disease in the Netherlands from 1870 to 1970," in *Veterinary Work in the Netherlands*, 161–65.

Initially, only diseased and suspected animals were killed. But from then on, sound animals on farms surrounding the infected area also were proactively killed to stop spreading of the virus.[34]

The annual vaccination approach was very successful, with no bovine cases occurring until 1991. Outbreaks among pigs, which were not vaccinated, occurred in 1961 and 1967. In spite of several outbreaks in the surrounding countries, the Netherlands remained free from the disease until March 2001. In that month, calves imported from Ireland introduced the disease into the Netherlands. That was exactly ten years after a nonvaccination policy was adopted by the EU member states. This nonvaccination policy was based on models developed by agricultural economists who claimed that it would be more cost-effective not to perform the annual vaccination and rely on early warning and quick cull and slaughter in the case of new outbreaks.[35] Ironically, this devastating FMD epizootic started in Great Britain, one of the countries very much in favor of nonvaccination.[36] Without immunity, Dutch livestock were susceptible to FMD, and within a couple of months, twenty-six farms were infected. A rigorous preventive cull-and-slaughter policy on almost three thousand farms surrounding the infected areas was started. More than 265,000 animals were killed: 93,000 bovines, 118,000 pigs, 35,000 sheep, 8,000 goats, and 11,000 other animals (e.g., deer). The economic damage amounted to some 2.8 billion Dutch guilders in total.[37] Veterinarians complained that they had to kill thousands of sound animals, even though new vaccines were available to solve the problem.[38] With these so-called DIVA[39] or marker vaccines, it is possible to differentiate between infected and vaccinated animals. Today, specialists

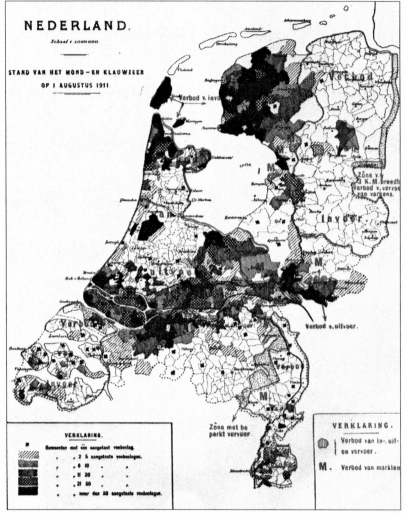

Figure 1.2. Spread of foot-and-mouth disease in the Netherlands in 1911. The dark areas represent the municipalities with the highest infection rate. The country is divided into various zones with bans on markets, imports, and exports. "Het Mond- en Klauwzeer in Nederland in 1911," *Verslagen en Mededeelingen van de Directie van den Landbouw*, 9, no. 1 (1912): following page 178

agree that a combination of cull and slaughter on infected farms and ring vaccination of animals on farms surrounding the infected area would be the best strategy.[40] This view is reflected in the EU directive (EG/2003/85) adjusted in December 2003.

While the impact of animal plagues over three centuries on the agricultural sector and on the development of veterinary medicine was relatively obvious, relatively less attention has been paid to the impact these diseases had on the wider public.

Changing Societal Context

Compared to previous outbreaks of epizootic diseases, outbreaks in the second half of the twentieth century occurred under different circumstances. Livestock production was strongly stimulated by the national government and, since 1964, also by the European Economic Community (EC). From that year on, several EC directives concerning livestock health and cross-border trade between member states became effective. In modern society, the spread of epizootic diseases is further stimulated by the global trade of animals and foods of animal origin and by worldwide tourism. From a veterinary point of view, there is a strong need to limit such movement. As argued by John Fisher, examples of outbreaks of epizootics in the past and present show that there seems to be a continuing contrast between open borders with free movement of animals and animal products required by international commercial treaties and the need to limit such movements from a veterinary point of view to prevent epizootic diseases from spreading. As early as 1863, during the first international veterinary congress in Hamburg, the English veterinarian John Gamgee proposed to turn the trade of live animals into a trade of cooled or frozen meat that could be more easily controlled than live animal trade.[41]

In the second half of the twentieth century, innovations in agriculture and animal-production methods such as scaling-up, specialization, cooperation, and mechanization, as well as crossbreeding and artificial insemination, had stimulated the development of mass production in the livestock sector of the Western world. This so-called factory farming of pigs and poultry and, to a far lesser extent, of cattle, changed from a solitary operating entity into a portion of the greater production chain from primary production to the consumer. From the 1980s on, both the livestock industry and the veterinary profession were faced with challenges posed by concerned consumers, animal-rights activists, and environmental and antimeat lobbyists.[42]

As viewed against this social-economic background, outbreaks of epizootics elicited a different response from society. Outbreaks of classical swine fever and FMD as well as the bovine spongiform encephalopathy problem were counted among the negative effects of intensive farming on animal welfare. Critics argued that these outbreaks clearly demonstrated that the socially beneficial limits to mass production in factory farming had been reached. Criticism within society and politics made it very clear to modern livestock producers that they have to address not only economic factors and international competition but also consumer and environmental protection, animal health and welfare, and national and international food-safety policies.[43]

Other reasons for the changing attitude toward major outbreaks of epizootic diseases are the decreased contribution of the rural economy to the national income and the damage these outbreaks do to the tourism industry. In addition, strong protests came from people who kept ruminants as hobby animals. Like the "factory-farmed" livestock, their animals were also subject to destruction in the cull-and-slaughter campaigns. Finally, pressure was put on the Dutch minister of agriculture when he announced that there was a possibility that ruminants such as deer and roe-deer in national parks had to be killed preventively, as well as domestic livestock.

Public Response

The public response to epizootic diseases differed considerably over the three centuries. At first, politicians and the media focused mainly on the economic consequences these diseases posed to agricultural policy. The majority of consumers were only interested in meat and milk prices; they did not care much about cull and slaughter or animal welfare. The impact of epizootics on farmers has always had both a financial and an emotional component as animals died or had to be killed. However, religion played a significant role in making it less difficult for farmers to accept their fate and move on. Moreover, the higher milk and meat prices during outbreaks enabled most farmers to survive the crisis.

By the late twentieth century, the changing human-animal relationship began to play a significant role in society's response to epizootic outbreaks. Recent outbreaks of swine fever and FMD have led to great societal commotion and criticism of intensive livestock farming. This criticism was extended to veterinarians who were accused of facilitating animal production without paying attention to animal welfare. Since these outbreaks, the veterinary profession has begun reappraising its role in factory farming. For instance, the fact that Dutch veterinarians were forced to kill thousands of sound animals to comply with EU directives met great opposition within the profession. The Netherlands Veterinary Medical Association was forced by its members to discuss this problem with the Dutch government. In April 2001, Dutch veterinarians for the first time in the history of their profession held a protest meeting in The Hague against the nonvaccination policy.[44]

Modern media, which had become less distant and more critical during the last decades, were responsible for putting epizootic diseases in the spotlight. Helicopters were used to get spectacular and dramatic images of how livestock were killed on the farms and transported to rendering plants. Broadcasted interviews often featured crying farmers' wives criticizing the cruelty of the government, which appeared to be more interested in strictly following EU regulations than in looking after the interests of common people. Agronomists and veterinarians, as well as representatives of several other so-called "specialist" groups, expressed their opinions on epizootic diseases on television. Intellectuals made comparisons to practices in Dachau and Auschwitz. During the epizootic, protest meetings were organized by several societal groups; unified groups with names like "Farmer and Civilian" were established, and some societies pleaded for immediate vaccination, which they saw as a less drastic solution than cull and slaughter. A special television program was organized to raise money for victimized farmers.[45]

Farmers complained that they were not fairly compensated for their losses, yet complaints sometimes came from the same farmers who viewed cessation of the annual preventive vaccinations against FMD in 1991 as a way to save money on veterinary expenses. The Dutch veterinary profession did not criticize the nonvaccination policy in 1991, waiting until 1995, when the Netherlands Veterinary Medical Association set up a committee to study that policy. By the end of that year, the committee concluded that,

Figure 1.3. Plea for immediate vaccination of cattle in Europe infected with foot-and-mouth disease in 2001. This image was on the cover of a postcard that could be sent to the European Union in Brussels. Courtesy of the action group Ent Europa, Kootstertille, Netherlands

with respect to highly infectious diseases such as FMD, a nonvaccination policy involved many risks and argued that the Dutch minister of agriculture and the EU commissioner for agriculture should reconsider the nonvaccination policy.[46]

In 1961, more than 320,000 pigs were killed to stop FMD, but the slaughter drew little media or popular attention. In 1997 and 2001, however, the systematic killing of large numbers of sound animals met great opposition throughout society. This public response clearly showed the large gap that had developed during the last decades between the modern urban consumer and producers living in the countryside. Modern factory-farming practices remained hidden to the broad public, since livestock production had moved far from urban life to behind the walls of stables and slaughterhouses. In short, the modern urban consumer became far removed from livestock production, knowing little but holding a romanticized view of happy farmers lovingly caring for their animals. Without the benefit of historical context, citizens regarded the recent outbreaks as a modern phenomenon caused by the high density of modern cattle stocks. Historians had to explain that epizootic diseases have always occurred throughout history.[47] This meant relatively little when changing public opinion led the government to reevaluate its policy toward intensive livestock production and livestock-disease control. Since the swine-fever outbreak of 1997, livestock numbers were significantly decreased by government-sponsored initiatives. Parties involved in livestock production began to realize that dealing with a largely uninformed public and media was almost more important than dealing with international trade interests and economic factors.

IN THE eighteenth century, state interference with cattle diseases remained limited. In the course of the nineteenth and twentieth centuries, though, a shift occurred from local and provincial control to national and international (EU) legislation regarding measures to deal with epizootic diseases. Some striking analogies among the responses to outbreaks of epizootic diseases can be observed with Dutch authorities' willingness to take drastic measures only when cattle exports decreased because foreign countries closed their borders. Progress in veterinary medicine was made subservient to the livestock economy. Veterinary logic continued a policy of limiting the free movement of animals to prevent outbreaks, supplementing that strategy with vaccination. Cull and slaughter continued to be a key element in the state's strategy of FMD and swine-fever control in the twentieth century. Although veterinary science developed rapidly after World War II, leading to more subtle ways of controlling animal diseases, international trade interests continued to dictate radical measures such as cull and slaughter. The public response to animal-disease crises has changed dramatically during the last decades of the twentieth century, leading to an emphasis on animal health and welfare. Media coverage of the recent outbreaks of swine fever and FMD in the Netherlands, coupled with a lack of knowledge of animal husbandry among a predominately urban population, led to changes in public opinion. The government responded by changing its policy with respect to livestock production, while the veterinary profession began reappraising its attitude toward production animals, just as it had done years before with companion animals. The responses of farmers, the meat trade, the meat and dairy industries, veterinarians, the media, and the wider public to outbreaks of epizootic diseases emphasize the importance that negotiation on all levels of everyday life has on legislation. Many radical measures concerning animal health that are firmly dictated by national and international trade interests, but not always closely followed, eventually turn out to be negotiable when critically discussed by the various actors in society.

Notes

1. Wilma Gijsbers, *Kapitale ossen: De internationale handel in slachtvee in Noordwest-Europa (1300–1750)* (Hilversum, Netherlands: Verloren, 1999), 83–95; Wilma Gijsbers and Peter A. Koolmees, "Food on Foot: Long-Distance Trade in Slaughter Oxen between Denmark and the Netherlands (Fourteenth–Eighteenth Century)," *Historia Medicinae Veterinariae* 26, nos. 3–4 (2001): 115–27; G. Steger, "Frischfleisch und rinderpest—Probleme frühere handelswegen," *Deutsche Tierärztliche Wochenschrift* 93, no. 4 (1986): 150–54.

2. Ronald Rommes, "'Geen vrolyk geloei der melkzware koeijen': Runderpest in Utrecht in de achttiende eeuw," *Jaarboek Oud-Utrecht* (2001): 87–135; idem, "Twee eeuwen runderpest in Nederland (1700–1900)," *Argos* 31 (2004): 33–40.

3. John R. Fisher, "The Economic Effects of Cattle Disease in Britain and Its Containment, 1860–1900," *Agricultural History* 54, no. 2 (1980): 278–94.

4. Jean Blancou, *History of the Surveillance and Control of Transmissible Animal Diseases* (Paris: Office International des Épizooties, 2003); Angela von den Driesch and Joris Peters, *Geschichte der Tiermedizin: 5000 Jahre Tiermedizin* (Stuttgart: Schattauer, 2003); F. C. Hekmeijer, *Korte geschiedenis der runderpest* (Amersfoort, Netherlands: Jacobs en Meijers, 1845).

5. In this chapter, the term *rinderpest* is used instead of *cattle plague* since the latter has had many different meanings in different places and at different times. Rinderpest hits all cloven-hoofed animals; here only rinderpest in bovines is discussed.

6. J. A. Faber, "Cattle-Plague in the Netherlands during the Eighteenth Century," *Mededelingen van de Landbouwhogeschool te Wageningen* 62, no. 11 (1962): 1–7; Steger, "Frischfleisch und rinderpest," 152–54.

7. Gijsbers and Koolmees, "Food on Foot," 123–24.

8. Ibid., 124.

9. Peter A. Koolmees, "The Role of Veterinary Medicine in the Development of Factory Farming," in *The Human-Animal Relationship: Forever and a Day*, ed. Francien de Jonge and Ruud van den Bos (Assen, Netherlands: Van Gorcum, 2005), 249–64.

10. Von den Driesch and Peters, *Geschichte der Tiermedizin*, 164–68; *The Merck Veterinary Manual*, http://www.merckvetmanual.com/mvm/index.jsp; Clive A. Spinage, *Cattle Plague: A History* (New York: Kluwer Academic, 2003), 15, 679; part 2 of this book is about the history of cattle plague in Europe.

11. J. A. Faber, "De veepest in Nederland in de achttiende eeuw," *Spiegel Historiael* 1 (1966): 67–74; Cees Offringa, *Van Gildestein naar Uithof: 150 jaar diergeneeskundig onderwijs in Utrecht* (Utrecht: Faculteit der Diergeneeskunde, 1971), 1:14–15.

12. Jan Bieleman, *Geschiedenis van de landbouw in Nederland, 1500–1950* (Meppel, Netherlands: Boom, 1992), 110–11, 161–66.

13. Rommes, "Twee eeuwen runderpest," 34.

14. Bieleman, *Geschiedenis van de landbouw,* 164; Rommes, "Twee eeuwen runderpest," 36.

15. A. van der Schaaf, "Geert Reinders (1737–1815), a Founder of the Practical Application of Immunology in the Fight against Infectious Diseases in Animals," *Historia Medicinae Veterinariae* 3, no. 4 (1978): 89–98.

16. Peter Verhoef, ed., *"Strictly Scientific and Practical Sense": A Century of the Central Veterinary Institute in the Netherlands, 1904–2004* (Rotterdam: Erasmus, 2007), 17.

17. Jan Willem Buisman, *Tussen vroomheid en Verlichting: Een cultuurhistorisch en -sociologisch onderzoek naar enkele aspecten van de Verlichting in Nederland (1755–1810)* (Zwolle, Netherlands: Waanders, 1992), 109–55; Constant Huygelen, "The Early Years of Vaccinology: Prophylactic Immunization in the Eighteenth and Nineteenth Centuries," *Sartoriana* 10 (1997): 79–110, esp. 85–88.

18. John R. Fisher, "Cattle Plagues Past and Present: The Mystery of Mad Cow Disease," *Journal of Contemporary History* 33, no. 2 (1998): 215–28; Offringa, *Van Gildestein naar Uithof,* 1:13–15.

19. Piet D. 't Hart, "Pestis bovina in Utrecht, 1813–1814," *Maandblad Oud-Utrecht* 46, no. 1 (1973): 2–6.

20. Gerardus J. Hengeveld, *Het rundvee: Zijne verschillende soorten, rassen en veredeling* (Haarlem: Erven Loosjes, 1872), 2:174–93, 227–55; Offringa, *Van Gildestein naar Uithof,* 1:111–13.

21. Hengeveld, *Het rundvee*, 240; Offringa, *Van Gildestein naar Uithof,* 1:112.

22. Bieleman, *Geschiedenis van de landbouw,* 291; C. J. Q. Kerstens, "How It Evolved," in *Veterinary Work in the Netherlands* (The Hague: Ministry of Agriculture, 1970), 12–25; Offringa, *Van Gildestein naar Uithof,* 1:114–15.

23. Von den Driesch and Peters, *Geschichte der Tiermedizin,* 169–70; Huygelen, "Early Years of Vaccinology," 94–97; *Merck Veterinary Manual,* http://www.merckvetmanual.com/mvm/index.jsp.

24. Hengeveld, *Het rundvee,* 103–7; Offringa, *Van Gildestein naar Uithof,* 1:66–69.

25. John R. Fisher, "To Kill or Not to Kill: The Eradication of Contagious Bovine Pleuro-pneumonia in Western Europe," *Medical History* 47, no. 3 (2003): 314–31; Constant Huygelen, "Louis Willems (1822–1907) and the Immunization against Contagious Bovine Pleuro-pneumonia: An Evaluation," *Verhandelingen van de Koninklijke Academie voor Geneeskunde van België* 59, no. 4 (1997): 237–85; Offringa, *Van Gildestein naar Uithof,* 1:109–11.

26. Peter A. Koolmees, *Symbolen van openbare hygiëne: Gemeentelijke slachthuizen in Nederland, 1795–1940* (Rotterdam: Erasmus, 1997), 70–75.

27. Von den Driesch and Peters, *Geschichte der Tiermedizin,* 181–82; *Merck Veterinary Manual,* http://www.merckvetmanual.com/mvm/index.jsp.

28. Verhoef, "Strictly Scientific and Practical Sense," 25–42.

29. K. G. Robijns, "Swine Fever and Swine Fever Eradication in the Netherlands," in *Veterinary Work in the Netherlands,* 174–85.

30. "Evaluatie voorkoming, opsporing en bestrijding klassiek varkenspest 1997–1998," *Tijdschrift voor Diergeneeskunde* 123, no. 18 (1998): 545–48; Aalt A. Dijkhuizen, "Het non-vaccinatiebeleid in economisch perspectief," *Tijdschrift voor Diergeneeskunde* 124, no. 3 (1999): 84–86; C. Terpstra, "Classical Swine Fever: From Serum Therapy to a Non-vaccination Strategy," in Verhoef, "Strictly Scientific and Practical Sense," 175–80.

31. H. J. Blokhuis et al., "Farm Animal Welfare Research in Interaction with Society," *Veterinary Quarterly* 22, no. 4 (2000): 217–22; Peter A. Koolmees, "From the Marshall Plan to Present-Day Prosperity: Veterinary Medicine in the Netherlands, 1945–2000," *Schweizer Archiv für Tierheilkunde* 144, no. 1 (2002): 24–31, esp. 25–26.

32. Von den Driesch and Peters, *Geschichte der Tiermedizin,* 170–73; *Merck Veterinary Manual,* http://www.merckvetmanual.com/mvm/index.jsp.

33. Offringa, *Van Gildestein naar Uithof,* 1:172–73; Dick J. Vervoorn, "Control of Foot-and-Mouth Disease in the Netherlands from 1870 to 1970," in *Veterinary Work in the Netherlands,* 155–65; *Verslag aan de Koningin van de bevindingen en handelingen van het Veeartsenijkundig Staatstoezicht in het jaar 1911* ('s Gravenhage: Van Langenhuysen, 1912), 70–71; "Het mond- en klauwzeer in Nederland

in 1911," *Verslagen en Mededeelingen van de Directie van den Landbouw* 9, no. 1 (1912): 1, 86.

34. Verhoef, *"Strictly Scientific and Practical Sense,"* 78–82; J. G. van Bekkum, "Dr. H. S. Frenkel, een coryfee uit de beginjaren van de mond- en klauwzeerbestrijding," *Tijdschrift voor Diergeneeskunde* 126, no. 10 (2001): 354–56; Offringa, *Van Gildestein naar Uithof: 150 Jaar diergeneeskundig Onderwijs in Utrecht* (Utrecht: Faculteit der Diergeneeskunde, 1981) 2:47–49, 192–93.

35. P. B. M. Berentsen, A. A. Dijkhuizen, and A. J. Oskam, "A Critique of Published Cost-benefit Analyses of Foot-and-Mouth Disease," *Preventive Veterinary Medicine* 12, nos. 3–4 (1992): 217–27; idem, "A Dynamic Model for Cost-Benefit Analyses of Foot-and-Mouth Disease Control Strategies," *Preventive Veterinary Medicine* 12, nos. 3–4 (1992): 229–43.

36. Abigail Woods, "The Construction of an Animal Plague: Foot and Mouth Disease in Nineteenth-Century Britain," *Social History of Medicine* 17, no. 1 (2004): 23–39.

37. Jaap de Boer, *Het ga je goed, lieve Evelien: MKZ-dagboek van een dierenarts* (Naarden: Strengholt, 2002), 10; *Dossier Diergezondheid—MKZ* (The Hague: Ministry of Agriculture, Nature and Food Quality, 2004).

38. De Boer, *Het ga je goed*, 13–17; Willem Schaftenaar, "Ik schaam mij als dierenarts," in *De toekomst van de boer*, ed. Peter Dekkers, Trouw dossier NL nr. 10 (Amsterdam: M. Muntinga b.v., 2001), 17–21.

39. The acronym stands for Differentiating Infected from Vaccinated Animals. See Terpstra, "Classical Swine Fever," in Verhoef, *"Strictly Scientific and Practical Sense,"* 179.

40. Jos P. T. M. Noordhuizen et al., "The Veterinarian and the Primary Production of Meat: Epidemiology and the Concern about Infectious Diseases and Risks," in *Veterinary Aspects of Meat Production, Processing, and Inspection: An Update of Recent Developments in Europe*, ed. Frans J. M. Smulders (Utrecht: ECCEAMST, 1999), 55–74.

41. John R. Fisher, "Professor Gamgee and the Farmers," *Veterinary History* 1, no. 2 (1979/80): 47–63; Fisher, "Cattle Plagues," 227; Richard Perren, *The Meat Trade in Britain, 1840–1914* (London: Routledge and Kegan Paul, 1978), 74–79, 100–5.

42. Susan D. Jones, *Valuing Animals: Veterinarians and Their Patients in Modern America* (Baltimore, MD: Johns Hopkins University Press, 2003), 91–114.

43. Koolmees, "The Role of Veterinary Medicine," 261–63.

44. Rens van Dobbenburgh, "Een dierenarts blijft een dierenarts," *Tijdschrift voor Diergeneeskunde* 126, no. 10 (2001): 364–65; Henk Vaarkamp and Sophie Deleu, "Dierenartsen protesteren tegen non-vaccinatiebeleid," *Tijdschrift voor Diergeneeskunde* 126, no. 10 (2001): 358–61.

45. Hans Siemens, *Ruimen: 10 keer erger dan ik dacht; MKZ-Het boerenverhaal* (Doetinchem, Netherlands: Elsevier, 2001), 60–86.

46. Susan Umans and Sophie Deleu, "MKZ-uitbraak noopt tot herbezinning," *Tijdschrift voor Diergeneeskunde* 126, no. 8 (2001): 300–301.

47. See, for instance, Jan Bieleman, "Van alle tijden," in Dekkers, *De toekomst van de boer,* 66–74.

Chapter 2

The Now-Opprobrious Title of "Horse Doctor"

Veterinarians and Professional Identity in Late Nineteenth-Century America

Ann N. Greene

AMERICAN veterinary history has enormous promise as a research field, due to its archival resources and conceptual potential. However, its secondary literature is problematic and frustrating. With few exceptions, it portrays veterinarians as scientific professionals of the modern state, heroically battling animal disease from their laboratories and protecting public health and the agricultural economy.[1] Written during the middle decades of the twentieth century as veterinarians struggled for status and identity, this literature positions veterinary medicine in the master narrative of scientific progress by modeling it after the prestigious fields of human medicine and bacteriology. However, paradigms from those fields elide more than they reveal of veterinarians' particular history.[2] This chapter suggests some ways that we might begin to write a new American veterinary history.

A curious feature of traditional veterinary history is that actual animals—material, historical animals—are largely absent. Instead, the focus is on various disease agents or the broad category of "livestock health," shifting the location of veterinary history away from field practice and into the more prestigious setting of the laboratory. Laboratories are supposed to be generic spaces with invariable tools and practices that produce universal knowledge, in contrast to work in the field. As historian Robert Kohler writes, "Laboratory science seems always to be granted a higher standing than field science. . . . It is precisely the stripped down simplicity and invariability of labs—their placelessness—that gives them their credibility." Field practice is "the result of a unique local history, never quite the same from one moment to the next, unpredictable, unrepeatable, beyond human control."[3] Veterinary medicine in the United States, which professionalized during the "golden age of bacteriology" that followed the introduction of germ theory in the 1870s, early hitched its wagon to the rising star of laboratory science, leaving a lacuna in veterinary history concerning the animals and human-animal relations of field practice.

Veterinary medicine by necessity encompasses both field practice and laboratory science. As a profession, it is situated between laboratory and field, a place Kohler calls "a zone of mixed practices and ambiguous identities" and "a place of mixed cultures, where . . . either side adopt each others' practices and develop approaches that are neither pure lab or pure field."[4] This is the terrain a new veterinary history should explore. Traditional histories disparage early practitioners as ignorant, low-class "horse doctors." This creates a historical problem, since, until at least the 1920s, the majority of veterinarians actually *were* horse doctors. As long as urban horses continued to provide the largest market for veterinarians, horse doctoring defined most of what veterinarians did, so much so that the decline in urban horse populations after 1915 created a significant crisis in the profession. However, veterinarians also treated other species. The equine-human relationship in veterinary medicine was different from the bovine-human, the swine-human, the canine-human, the elephant- (and other zoo animals) human, or the disease agent- ("germ") human relationship. A new history of veterinary medicine must examine both field practice and laboratory science and place the historicity and specificity of animals and human-animal relations at the center of its concern. It must start describing actual practice.[5]

How should one write such a history? The way to explore the terrain of "mixed practices and ambiguous identities" that constitutes veterinary history is by using the concept of ecology as a method of historical inquiry. Ecology considers specific communities of organisms, their internal relationships, and their interactions with their surroundings.[6] Charles Elton, one of the founders of ecology, called it "scientific natural history" and emphasized attention to the various niches, or functions, of the members of an ecosystem. Because ecology emphasizes place and relationship, it provides a way for veterinary historians to bring field sites, laboratories, and human-animal relations together. Because ecology is attentive to change but not concerned with progress, the ecosystem concept avoids the problem of progress inherent in the analytical paradigm of professionalization. An ecological approach considers how ideas take form in specific places— in this case, how scientific ideas and altered concepts of diseases and therapeutics entered nineteenth-century society, the extent to which they altered perception and practice, and the characteristics and inhabitants of the sites where this process occurred.[7] Finally, an ecological approach can bring the coevolution of humans and animals, how humans and animals shape each other, into veterinary history.[8]

In this chapter, I use this approach to examine the early history of the School of Veterinary Medicine at the University of Pennsylvania (Penn) in Philadelphia and to explore the development of veterinary knowledge in the late nineteenth century. Penn's veterinary school is significant in the history of veterinary medicine because it is one of the few schools established in the nineteenth century that survived the crisis in veterinary medicine that occurred after 1915. Forty-eight veterinary schools and programs opened between 1850 and 1915 in the United States. Of these, forty closed their doors by 1920. Between 1914 and 1924, the number of veterinary students declined by 75 percent. Many Americans assumed that the veterinary profession would disappear as motorization took over the technological niches that had been occupied by horses. By the mid-1920s, only Penn and a handful other schools were left, their deans trying to convince people that veterinarians were not doomed to extinction.[9] Because Penn's veterinary school's history is coterminous with the history of the American veterinary profession, it reflects the many professional, institutional, and cultural issues facing veterinarians in late nineteenth-century and early twentieth-century America. Through the early history of the School of Veterinary Medicine, it is possible to see the connection between the development of veterinary medicine and central questions in national and transnational history about state formation, institutional development, industrialization, and environmental change.

American veterinary medicine developed as an industrial profession because industrialization altered the demography of animal populations and the ecology of human-animal relations in nineteenth-century America. Livestock populations expanded to support a growing population, and the population of work animals, most of which were horses, grew dramatically to meet the rising consumption of motive power. Though many histories state that the "iron horse" of the railroads replaced real horses, the opposite occurred. Horses supplied local, flexible-route transport, while railroads provided long-distance, fixed-route, high-volume transport. This complementary relationship between iron and real horses caused equine populations to burgeon along with the railroad network. In addition, mechanized agriculture relied almost exclusively on horse power until the end of the century. The American horse population multiplied nearly five-fold in the second half of the nineteenth century, compared to a three-fold increase for the human population. Whereas there was one horse for every five or six humans in 1850, by 1900 there was one horse for every three humans, on average.[10]

Industrialization made possible concentrated populations of animals, such as army horses during the Civil War, cattle herds on western ranges, and the animal residents of cities. The urban setting of the University of Pennsylvania made it a good site for a veterinary school. Schools in rural areas (such as Cornell and Iowa) were not considered good locations because there would not be enough animals to support veterinary services. Like other nineteenth-century industrial cities, Philadelphia was an "organic city" that teemed with creatures great and small. Human residents encountered cattle and sheep driven through the streets, packs of roving pigs, feral cats and dogs, pet cats and dogs, an invasion of English sparrows, a variety of poultry, verminous rodents and insects, and, though invisible to them, an array of microorganisms.[11] Most of all, they encountered horses, as ubiquitous in Gilded Age cities as squirrels in today's. According to historian Philip Teigen, a person in a city like Philadelphia was likely to encounter more horses than a cowboy in Texas.[12] Horses and humans were urban coworkers and coresidents, and encounters between them were intimate and immediate. Horses are large, social animals whose big eyes meet humans' at eye level; they often have individual names and identities as well. After five thousand years of coevolution and codomestication, horses and humans have developed a special relationship.

In the United States, horses urbanized 50 percent faster than humans between 1870 and 1900. Urban human populations rose 219 percent, while the urban horse population grew 371 percent. By 1900, Philadelphia had nearly four hundred horses per square mile. During the late nineteenth century, Philadelphia housed thousands of horses, of which five thousand powered its mass transit system; the rest worked in hauling, delivery, transportation, construction, and manufacturing.[13] Horses were part of the function or focus of many of the economic, political, and civic institutions of city life. Philadelphia had consolidated in 1854 and since had developed an extensive government bureaucracy. Police, fire, and sanitation services all used horses. Philadelphia also was home to a plethora of civic institutions of science, education, and reform, one of which was the University of

Pennsylvania. In 1869, Philadelphians established the Pennsylvania Society for the Prevention of Cruelty to Animals (PSPCA) with a large, upper-class membership for whom horses and animal welfare were important parts of their cultural and class identity and who were potential consumers of and advocates for veterinary services.

In the 1870s, the University of Pennsylvania moved from Center City across the Schuylkill River to West Philadelphia.[14] The appointment of medical professor William Pepper as provost in 1881 opened a new chapter in Penn's development. Penn's prestigious medical school dated from 1765, enjoyed a national reputation, and overshadowed the rest of the university, which remained a rather parochial institution. Pepper was a prominent physician from the Philadelphia elite, a brilliant and forceful leader who "drove the power of his mind as he drove his horses through the streets" from his residence on fashionable Rittenhouse Square to the university.[15] As a member of the medical faculty, Pepper was responsible for founding and raising funds for the first university hospital in the United States—the first new building on the West Philadelphia campus. He was determined to increase the rigor of medical training and raise professional standards; at Penn, he accomplished a revised curriculum, tougher requirements for admission and graduation, and salaries for medical professors. As the first provost empowered by the board of trustees to be chief administrative officer, he expanded his vision of educational reform to the entire university. During his seventeen-year tenure, Pepper added sixteen departments and schools, including the veterinary school, the Wharton School, and the School of Fine Arts.[16] From the 1870s on, with ongoing construction amid streets filled with ordinary horse-drawn traffic, the university had horses everywhere.

There were soon to be more. In the 1870s, Horatio J. Smith proposed that Penn offer courses in veterinary medicine. The secretary of the National Agricultural Congress, Smith managed a large horse farm in West Philadelphia. In his experience, it was difficult to obtain competent medical care for animals even on the fringe of the city, and he knew that farmers in rural areas lacked access to any veterinary services. Smith was also concerned about animal epidemics. Disease agents were another burgeoning population in industrial America. The circulation of large numbers of animals through national and international markets and the large, concentrated animal populations in stockyards and cities created unprecedented opportunities for contagion and epidemics. Outbreaks of cattle pleuropneumonia, hog cholera, glanders, Texas cattle fever, and horse influenza caused economic losses, disrupted national and international trade in animals and animal products, jeopardized human health through contaminated meat and milk, and impaired city services.

In an "earnest plea" read before the Pennsylvania Board of Agriculture, Smith suggested a program of "co-education" that supplemented medical education with veterinary training: "The addition of a year, perhaps less, to the curriculum of our doctors, would fit them to practice veterinary medicine; and if the absurd and injurious prejudice against the andropath acting as a zoopath were broken down we could in a very short time have thousands of competent veterinary practitioners," especially in rural districts. Smith circulated his plea widely among veterinarians, doctors, and the agricultural community.[17] In general, veterinarians saw coeducation as a threat because they thought medical doctors with veterinary training would take away their practices. Doctors, on the other hand, thought that treating animals would diminish their status.[18] Pepper, fresh from the battle to revise the medical-school curriculum, favored establishing a separate veterinary department. Penn's board of trustees approved the establishment of the veterinary department early in 1878, contingent on raising enough money to fund it.[19]

One of Pepper's accomplishments as provost was getting the famously parsimonious Philadelphia elite to give money to the university, and some of his first successes were with donations for the veterinary school.[20] In 1882, board member J. B. Lippincott, a successful Philadelphia publisher and president of the Philadelphia branch of the SPCA, gave ten thousand dollars toward endowing a Veterinary Department. Fairman Rogers, another board member, donated five thousand dollars. Rogers, a Penn engineering professor from 1855 to 1871, was a founding member of the National Academy of Science, president of the Academy of Fine Arts, and a horseman of international reputation. Joseph E. Gillingham, an entrepreneur and civic leader, donated ten thousand dollars. With this money in hand, the board authorized construction of buildings on a triangular piece of land at Thirty-sixth and Pine streets on the south edge of the university campus.[21]

In 1883, the Veterinary Committee of the board selected Rush Shippen Huidekoper to be the first dean of the school and also to serve as professor of veterinary anatomy and physiology. Huidekoper's name revealed his membership in the old Philadelphian Rush and Shippen families. He was related to renowned Penn physicians Benjamin Rush and William Shippen; to Edward Shippen, the first mayor of Philadelphia; and to Henry Shippen Huidekoper, a Pennsylvania Civil War hero. In addition to being well connected socially, he was known in elite circles as a superior horseman and was a member of the elite Rose Hill Hunt Club with Fairman

Rogers. Huidekoper had a medical degree from Penn, but on Rogers's suggestion went to France to earn a veterinary degree from the famous school at Alfort, with the understanding that he would be considered for dean of the veterinary school when he returned. It was not unusual for people to go to Europe for training, due to the lack of veterinary schools in the United States. Like many other Americans receiving European training, Huidekoper worked with Louis Pasteur after graduating and also visited the laboratories of Robert Koch and Rudolf Virchow to give his veterinary education an extra bacteriological shine. On his return to Philadelphia, Huidekoper was enthusiastically recommended to the board for the dean position by Fairman Rogers, who wrote, "There is no one who can come in competition with him," adding that Huidekoper was "a horseman of the best quality and passionately devoted to the subject."[22] Together, Smith, Lippincott, Gillingham, Rogers, Huidekoper, and Pepper constituted a unique group, all elite Philadelphians bound together by social and class ties and by their interests in horses, veterinary medicine, science, photography and art, animal welfare, and the university. This group, and the social milieu they represented, was a critical element of the school's founding.

The School of Veterinary Medicine opened in 1884. At its new facility, wide gates opened from the street into a spacious yard that could accommodate wagons, carriages, ambulances, and a number of animal patients. Examination rooms, stabling for ill horses, the shoeing forge, and buildings for classrooms, offices, laboratories, and the library ringed the central yard. Teaching reflected Huidekoper's desire to pattern the curriculum after his own European training. He established rigorous standards from the beginning and emphasized both laboratory science and field practice in the school's curriculum. Entrance requirements for the veterinary school were the same as those of the medical school, and first-year students shared core classes with medical and dental students. Horses were the primary patients of the school, and veterinary students spent two afternoons a week in the forge learning how to assess the condition of a horse's hooves and to shoe properly, so that in the future they could diagnose foot ailments and supervise corrective shoeing. Students were expected to spend all their spare time in the dissection rooms, hospital wards, and the forge. Twelve out of twenty-nine students failed to pass the first year. Those that survived moved on to the second-year and third-year curriculum that included pathology, therapeutics, anatomy, and contagious diseases.[23]

Huidekoper's plans to expand the school's programs and facilities illuminate the unique characteristics of animal doctoring, as well as his vision of human-animal relations at the school. He established an ambulance service and noted that it trained students "to handle carefully and properly sick and injured animals, which as the case of a horse weighing 1,600 or 1,800 pounds and suffering intensely, is sometimes extremely difficult." He established a shoeing shop to bring in patients, provide clinical training, and generate income. Huidekoper wanted to add a cattle stable and a dog kennel and to build dormitories so that each student could be "in direct personal contact with the animal he treats. . . . It is a most arduous labor to familiarize our students (many of whom are from cities) with even the normal conditions of cattle and swine for which we need the proper stables on the grounds."[24]

The school participated in a unique form of animal research when William Pepper and Fairman Rogers brought photographer Eadweard Muybridge to the university in 1884 to undertake an extensive photographic study of human and animal locomotion. In the late 1870s, Rogers became interested in Muybridge's work and commissioned Philadelphia artist Thomas Eakins, who was already using photography in his work, to produce a painting of Rogers's coach horses in action, *A May Morning in the Park*, using Muybridge's techniques. Pepper commented, "In a larger conception of [the university's] duty should be included the aid which it can extend to investigators engaged in researches too costly or elaborate to be accomplished by private means."[25] In the late 1870s, Muybridge had revolutionized photography and settled a centuries-old debate when his stop-action photographs of a horse in motion proved that there is a point in the gallop when all four feet are off the ground. By capturing motion on film, he abstracted motion from the physical body, turning motion into generic laboratory knowledge rather than field observation. Between 1884 and 1897, Muybridge worked at the veterinary school, producing hundreds of photographs of humans and animals in motion, later published as *Animal Locomotion* and *Human Locomotion*. He photographed a wide range of domestic animals and went to the Philadelphia Zoo to photograph wild animals. He devoted one entire volume to horses, photographing their movements walking, trotting, galloping, jumping, and pulling. Muybridge also photographed injured horses, to assist in understanding and diagnosing lameness. Huidekoper appeared in some of Muybridge's photographs on his favorite mare Pandora—in some photographs fully clothed and in others discreetly nude.[26]

Photography was one of the technologies transforming Americans' knowledge of the world as they became self-consciously modern, and horse photography was intertwined with other aspects of nineteenth-century society. It was already changing how people saw and represented the world. In medicine, both human doctors and veterinarians studied Muybridge's photographs to understand changes in movement caused by human and animal deformity and injury. Muybridge's initial work on horse locomotion was sponsored by Leland Stanford, a self-made man and railroad magnate who was not only interested in his racehorses but in the industrial applications of Muybridge's technique. The abstraction of motion in Muybridge's work and that of Étienne-Jules Marey in France contributed to the rationalization of labor systematized by Frederick Winslow Taylor, to the motion studies of Frank and Lillian Gilbreth, and to industrial management practices.

In 1887, the school graduated its first ten students, conferring on them the degree of Veterinary Medical Doctor (V.M.D.). Huidekoper observed with satisfaction,

> The small percentage of the students who had matriculated at the outset with the idea that a course of university study was a sinecure, which was to secure to them simply the protecting garb of the University diploma after the payment of a few fees, learned that the superiority which they were to obtain over the empiric practitioners of their neighborhood, could only be obtained by close attention, constant industry, and active intelligence. This small percentage dropped from the rolls either by their own volition or as the result of a rigid examination.[27]

Though Huidekoper used the term "empiric" as a straw man, the curriculum combined the empirical or experiential tradition of horse care with an emphasis on laboratory science. The name Penn selected for its degree—Veterinary Medical Doctor—reflected the relationship between the veterinary school and the medical school.

However, despite its apparent success, the school's existence was fragile. First, the circle of men instrumental in founding and funding the school began to disperse. Fairman Rogers moved abroad, leaving nearly one thousand horse books to the veterinary library. Lippincott died in 1884. Pepper's attention was divided among the medical school and the other schools and departments of the university. There is no record of whether Smith and Gillingham remained active supporters. Second, the veterinary school was in constant financial trouble and ran an annual deficit. The generous initial donations had gone toward constructing the school. Lippincott's heirs gave an annual donation of four thousand dollars, but the family complained that the school was expecting them to support it and resisted making a major gift toward endowment. There is no record that Rogers and Gillingham donated more money, and the mysterious Smith, even if he was still active in supporting the school, would not have had the financial resources to provide substantial funds. A subscription fund sponsored by the Philadelphia SPCA proved unsuccessful.

Huidekoper and the board hoped that fees from providing veterinary services would offset the expenses of running the school. However, despite its location in an urban environment with a large population of working horses and its founders' social connections with the horse-owning Philadelphia elite, the veterinary school struggled to establish itself in the city's horse economy. The shoeing shop lost money and could not seem to build up a steady clientele. The shop did not pay the customary 10 percent tip to coachmen who brought their horses to Penn. Therefore, coachmen took their horses elsewhere unless their employer specifically ordered them to go to Penn. Huidekoper pleaded with the board members to ask their rich horse-owner friends to send their horses to Penn for shoeing.[28] This vignette reveals the importance of the social environment in the ecology of the Penn veterinary school. Even among the elite horse owners of Philadelphia, decisions about shoeing—which affects a horse's health and soundness—were shaped by the culture of domestic servants and employers and by the gratuity system between coachmen and those providing equine services and goods, not by owners' decisions to seek the most consistent, scientific shoeing for their horses.

Huidekoper not only began using his own money to cover expenses but found that the board balked at reimbursing him. This and the school's chronic financial woes aggravated another sore point for Huidekoper—neither he nor the faculty received any compensation for their work. The board expected that the private practices of veterinary faculty would flourish because of the veterinarians' association with the school and that this additional income earned from their Penn affiliation would compensate them for teaching. This was the system traditionally used for medical school faculty at Penn and elsewhere, and the board expected it to work for the veterinary school as well. It was a practice that was being rendered extinct by the rise of the modern research university and graduate school and its accompanying demands for expertise and for administration. Huidekoper, with a demanding teaching schedule and full responsibility for administration and fundraising found it difficult to maintain his private practice.

Huidekoper also worried that, without paying salaries, Penn would not be able to recruit or retain faculty. New salaried job opportunities for veterinarians were appearing in the public sector. In the late 1860s,

Congress began appropriating money to study and control animal disease after an outbreak of Texas cattle fever led states to patrol their borders with armed posses and caused cattle ranchers to call for federal assistance. In 1884, Congress established the Bureau of Animal Industry and gave it research funds and regulatory authority. At the same time, state governments were establishing public health departments and regulating animal health, products, and travel. Veterinarians claimed this aspect of public health as their purview, and the public sector soon became a major source of employment for veterinary-school graduates.[29]

It is not clear from university records why the board of trustees did not respond to Huidekoper's concerns and provide more funding. Perhaps the founding of the veterinary school had been too much the project of a few trustees and lacked support from the board as a whole when those trustees were not longer there. The board consisted of an insular group of elite Philadelphians who for years had governed the university as a committee and only recently had empowered a provost (and Pepper happened to be one of their own) to be the chief administrative officer. Their understanding of the changing university environment in the late nineteenth century and the accompanying needs for funding and facilities lagged behind the thinking of the progressive, ambitious Huidekoper. Assuming that the school should be self-supporting, they perhaps blamed Huidekoper for the school's chronic budgetary woes.

Huidekoper turned to the public sector for financial support. In 1887, the school petitioned the state legislature for an appropriation of one hundred thousand dollars. Huidekoper worked his social and political connections to lobby for the bill, but the legislature approved an appropriation of only fifty thousand dollars. However, this was promptly vetoed by Governor Robert Pattison, a Democrat who had recently won office on a platform of government reform and economy. Pattison was trying to reduce the state debt by three million dollars. Since the university was in Philadelphia, a city controlled by Pattison's rival, Republican boss Matthew Quay, and had a largely Republican board of trustees, the political environment was perhaps not favorable for an appropriation for the veterinary school. In 1889, when Quay's candidate James A. Beaver became governor, Beaver approved an appropriation of twenty-five thousand dollars and twelve state-funded scholarships. But by then, Huidekoper, his relationship with Pepper having grown increasingly acrimonious, resigned as dean early in 1889 but remained on the faculty. In the following October, for reasons unclear, Huidekoper abruptly quit his faculty position at the veterinary school. He eventually left Philadelphia and moved to New York, was later president of the American Veterinary Medical Association, and worked to establish an Army Veterinary Corps.[30]

The difficulties experienced by the Penn veterinary school might also have reflected a more general indifference to veterinary education in the United States. Neither inadequate horse care nor recurring animal plagues seem to have been perceived as enough of a problem by enough Americans to inspire a broad commitment to veterinary medicine and science until well into the twentieth century. Hundreds of thousands of horse casualties in the Civil War did little to jump-start veterinary awareness or education. An epizootic of equine influenza in the early 1870s virtually shut down the urban Northeast, swept across the country, and contributed to an economic depression, yet successful, institutionally based veterinary schools did not appear for another decade. Livestock owners suffered costly losses from animal epidemics from mid-century on and demanded government help, yet resented and resisted attempts by the Bureau of Animal Industry to regulate livestock production and trade when they thought it would hurt them economically. The difficulties that Huidekoper encountered were not entirely due to the particular environment at Penn. Broad acceptance of a modern veterinary regime would take decades. Factors such as the expansion of the modern liberal state, the political economy of agriculture, public health concerns, and the changing cultural value not only of animals but of which animals were valued shaped the veterinary profession.

In the wake of Huidekoper's departure, state support enabled the School of Veterinary Medicine to survive its own financial troubles and later to weather the severe national depression of the 1890s. Leonard Pearson, a graduate of the veterinary school and State Veterinarian of Pennsylvania, became dean of the veterinary school in 1897. He moved the laboratory of the Pennsylvania State Livestock Sanitation Board to Penn. This and other kinds of state support enabled the veterinary school to survive the precipitous decline in the urban horse population and in the number of veterinary students after 1915.

At Penn, a school for horse doctors established itself successfully on the laboratory-field boundary. In the beginning, it had the blessing of the medical faculty, university leadership, the determination of an unusual group of horsemen-scientists, and an excellent location in a large industrial city. Later, it benefited from what became paradigmatic for veterinary education—the necessity of reaching beyond the borders of the university to draw on the expanding powers of the liberal state for financial support and professional legitimacy. Scientific medicine did not trump traditional horse medicine; the knowledge from the dissecting and bacteriology laboratories complemented the oral traditions and direct experience of traditional horse culture and medical practice in the field (and street and barn and stable).

Beginning in the central yard of the veterinary school, an ecological mode of inquiry reveals the complex web of relationships that constitutes the laboratory-field terrain of "mixed practices and ambiguous identities" at Penn. Veterinary history is usually a story of science's triumph that at Penn has emphasized Huidekoper and Pearson. A new veterinary history must see past what veterinarians wrote about professional knowledge and practice and discern actual practices and conditions. The story of Penn is a particular rather than a universal story. Yes, Huidekoper had a degree from Alfort, Fairman Rogers was a nationally known engineer, and William Pepper a leading physician. But Penn developed not only out of a traditional, elite horse culture; its history is a very Philadelphia story. It may be that a new veterinary history must be built up out of local stories, using methodologies drawn from social, cultural, environmental, and technological histories that examine context and contingency. Perhaps veterinary medicine developed less as a profession and more as groups of practitioners in a variety of centers; if there is an overarching historical narrative, it should come out of an array of histories, rather than vice versa.

Animals and animal-human relations rather than veterinarians should be central to veterinary history. Traditional history emphasized science because animals were not seen as serious objects of study. Yet the histories of animals and humans are intertwined, as shown by the centrality of horses to nineteenth-century society and the growing field of animal history and animal studies today. The very presence of horses and other animals makes them agents in history. As organic beings, they consume food, space, and artifacts and have a material presence; as sentient beings, they have a social and cultural presence; they are historical beings that have changed over time. Furthermore, unlike human doctors, veterinarians treat an array of different species, each with different physiologies, anatomies, immunologies, life cycles, and characters, and each of which has a different history with humans.

American veterinary history requires a transnational perspective as well. Huidekoper trained in Europe, and other veterinarians important in the United States at the time, such as James Law at Cornell University and Alexander Liautard in New York City, came from abroad. All tried to transplant a European veterinary regime to the United States. At the same time, imperial expansion imposed European values on traditional cultures and practices. The United States had its own imperial history in the conquest and development of the American West. Veterinary medicine benefited from the process of state building even in countries that developed veterinary regimes earlier than the United States. What different regimes emerged in different places? What is the relationship between industrialization and the development of veterinary medicine? How did the global expansion of capitalism, with its powerful forces of rationalization and commodification, affect the economic and cultural value of different animals and the political economy of veterinary medicine? Disease respects no political boundaries; there are questions to consider of domestic politics and foreign relations as disease agents move between wild and domestic populations, species, regions, and countries. Finally, there are questions of relations between veterinarians from different countries and between veterinarians and other professions, such as wildlife managers, conservationists, environmentalists, and ethologists.

This chapter has looked at the ecology of veterinary medicine as it emerged at the University of Pennsylvania School of Veterinary Medicine in the late nineteenth century—a school devoted largely to equine medicine. Rush Shippen Huidekoper, its guiding spirit for the first years of its existence, had both medical training and the most advanced veterinary training available at that time. He was a "scientific" veterinarian, not considered a "horse doctor." But Huidekoper was a bridge between the culture of traditional equine medicine, with its oral tradition, and the new culture of laboratory science. He was the heir to an old tradition even as he participated in its transformation. Though American veterinary history traditionally sought to distance veterinarians from the opprobrium of being known as horse doctors, it is time to bring the horse doctors back into veterinary history, beginning with professionals such as Huidekoper and his counterparts at other schools of the time and going back earlier to discover the rich tradition of horse doctoring that they inherited and expanded. In a new veterinary history, "the end of all our exploring / Will be to arrive where we started,"[31] by studying the no-longer opprobrious "horse doctors."

Notes

1. For examples of this kind of celebratory American veterinary history, see J. F. Smithcors, *The American Veterinary Profession: Its Background and Development* (Ames: Iowa State University Press, 1963); B. W. Bierer, *American Veterinary History* (1940; repr., Carl Olson, 1980), and idem, *A Short History of Veterinary Medicine* (East Lansing: Michigan State University Press, 1955); Fred Wilbur Powell, *The Bureau of Animal Industry: Its History, Activities and Organization* (Baltimore, MD: Johns Hopkins Press, 1927). These are earnest and well-intentioned histories, and their authors did important work in assembling

the material. However, they can be downright awful to read and difficult to extract information from. Two significant examples of new veterinary history are Susan D. Jones, *Valuing Animals: Veterinarians and Their Patients in Modern America* (Baltimore, MD: Johns Hopkins University Press, 2003), and the articles of Philip Teigen in *Veterinary History* and elsewhere.

2. Charles Rosenberg, "Toward an Ecology of Knowledge: On Discipline, Context and History," in *The Organization of Knowledge in Modern America, 1860–1920*, ed. Alexandra Oleson and John Voss (Baltimore, MD: Johns Hopkins University Press, 1979), 442.

3. Robert E. Kohler, *Landscapes and Labscapes* (Chicago: University of Chicago Press, 2002), 6–7.

4. Ibid., 18–19.

5. Edwin H. Ackerknecht, "A Plea for a 'Behaviorist' Approach in Writing the History of Medicine," *Journal of the History of Medicine and Allied Sciences* 22, no. 3 (1967): 211–14.

6. Charles Elton, *Animal Ecology* (1927; repr., London: Sedgwick and Jackson Ltd, 1951); *Encyclopedia of Ecology and Environmental Science*, ed. Peter Calow (Oxford: Blackwell Science, 1998).

7. Rosenberg, "Toward an Ecology of Knowledge," 441, 452.

8. Edmund Russell, "Evolutionary History: Prospectus for a New Field," *Environmental History* 8, no. 2 (2003): 209, 219.

9. Jones, *Valuing Animals*, 48–49.

10. U.S. Census Office, *Compendium of the Seventh Census* (Washington, D.C.: Beverly Tucker, 1851); U.S. Census Office, *Twelfth Census*, vol. 5 (Washington, D.C.: U.S. Census Office, 1902).

11. Ted Steinberg, *Down to Earth: Nature's Role in American History* (New York: Oxford University Press, 2002), 157; Robin Doughty, "The English Sparrow in the American Landscape," Research Paper no. 19, School of Geography, University of Oxford, 1978.

12. Philip Teigen, "Urban and Rural Horses and Mules in the United States, 1860–1920," working paper, National Institute of Health, Bethesda, MD, May 2001, 7.

13. Clay McShane and Joel Tarr, "The Centrality of the Horse in the Nineteenth-Century American City," in *The Making of Urban America*, ed. Raymond Mohl (Wilmington, DE: SR Books, 1997), 107. Equine population-density figures courtesy of Philip M. Teigen, National Library of Medicine, Bethesda, MD.

14. Chronological information about the history of the University of Pennsylvania and the School of Veterinary Medicine comes from *History of the School of Veterinary Medicine at the University of Pennsylvania, 1884–1934* (Philadelphia: Veterinary Alumni Society, 1935); John E. Martin, *A Legacy and*

a Promise: The First One Hundred Years of the School of Veterinary Medicine (Philadelphia: University of Pennsylvania, 1984).

15. *American National Biography* (New York: Oxford University Press, 1999), s.v. "Pepper, William, Junior"; *Addresses . . . in Memory of William Pepper* (reprinted for American Philosophical Society Memorial Volume, 1899), 37.

16. William Pepper, "Higher Medical Education, the True Interest of the Public and of the Profession: Two Addresses . . ." (Philadelphia: J. B. Lippincott, 1894).

17. Horace J. Smith, "Veterinary Science," UPC 5, box 6, FF 17, University of Pennsylvania Archives.

18. C. B. Michener to Horace J. Smith, 15 February 1878, UPC 5, box 6, FF 18; James Tyson to Horatio J. Smith, 10 December 1877, UPC 5, box 6, FF 19, University of Pennsylvania Archives.

19. Smith, who receives credit in all histories of the veterinary school for promoting the idea of such a school, disappears entirely from the school's archives after this point.

20. E. Digby Baltzell, *Philadelphia Gentlemen* (1958; reprint New Brunswick, NJ: Transaction Publishers, 1995), 322–25.

21. Claire Gilbride Fox, "Fairman Rogers: Professor, Trustee and Friend," in *The Fairman Rogers Collection on the Horse and Equitation: A History and Guide* (Philadelphia: School of Veterinary Medicine, University of Pennsylvania, 1975), 1–7; *F. R., 1833–1900* (Philadelphia: privately printed, 1903); UPA 6.2Gil; UPC 5, box 6, FF 3, University of Pennsylvania Archives.

22. Fairman Rogers to Board of Trustees, 29 March 1883, UPA 3, 1883—Veterinary, University of Pennsylvania Archives.

23. Photograph, William Rau, "Veterinary Department Courtyard, Original Building," 1902, UPX 12, box 39, FF 18; Minutes, Board of Trustees, 5 December 1882, UPC 5, box 6, FF 23; UPC 5, box 6, FF 13; Curricula, UPA 3, 1888—Veterinary, University of Pennsylvania Archives.

24. Rush S. Huidekoper to Board of Trustees, Report for 1886–87, UPA 3, 1887—Veterinary, University of Pennsylvania Archives.

25. William Pepper, "Note," in *Animal Locomotion: The Muybridge Work at the University of Pennsylvania* (New York: Arno, 1973), 5; Eadweard Muybridge, *Muybridge's Complete Human and Animal Locomotion*, vol. 3 (New York: Dover, 1979).

26. Siegfried Giedion, *Mechanization Takes Command* (New York: Oxford University Press, 1948), 17–25; Rebecca Solnit, *River of Shadows: Eadweard Muybridge and the Technological Wild West* (New York: Penguin Books, 2003), 185, 195, 219–22. In order to honor Pandora when she died, Huidekoper had her butchered, hosted an elaborate dinner in her memory, and informed his guests at the end of the evening that they had just eaten Pandora.

27. UPA 3, 1887—Veterinary, University of Pennsylvania Archives.

28. Rush S. Huidekoper to Board of Trustees, 31 August 1889, UPA 3, 1889—Veterinary; Craig Lippincott to William Pepper, 14 April 1888, UPA 3, 1887—Veterinary; Rush S. Huidekoper to Board of Trustees, 31 May 1886, UPA 3, 1886—Veterinary, University of Pennsylvania Archives.

29. Rush S. Huidekoper to Board of Trustees, Report for 1886–87, UPA 3, 1887—Veterinary, University of Pennsylvania Archives. See Nancy Tomes, *The Gospel of Germs* (Cambridge, MA: Harvard University Press, 1998), for changes in American ideas about disease and sanitary practices beginning in the 1870s.

30. Report for 1886–87, UPA 3, 1887—Veterinary; Rush S. Huidekoper to William Pepper, 7 January 1889, UPA 6.2Pep, FF 12, University of Pennsylvania Archives; Martin, *Legacy and Promise*, 37, 45; Wayland F. Dunaway, *A History of Pennsylvania*, 2d ed. (Englewood Cliffs, NJ: Prentice Hall, 1948), 448–54. Huidekoper (1854–1901) helped in the aftermath of the Johnstown flood as part of the militia. He later served in the Spanish-American War.

31. T. S. Eliot, "Little Gidding," in *Four Quartets* (New York: Harcourt Brace Jovanovich, 1971).

CHAPTER 3

Breeding Cows, Maximizing Milk

British Veterinarians and the Livestock Economy, 1930–50

Abigail Woods

WORLD WAR II precipitated dramatic changes in British agriculture, as enemy action and the need to preserve scarce shipping space undermined the nation's traditional reliance on food imports. Formerly a marginal industry that had struggled for economic survival throughout the interwar depression, agriculture became central to the health, strength, and fighting capacity of the nation. Under the direction of the state, the prevailing "low input–low output" approach was replaced by a drive for production at almost any cost. Milk was central to this campaign. Interwar advances in nutritional science had designated it a "protective food" essential for health.[1] Moreover, there was considerable capacity for its production within Great Britain. Interwar dairy farming had proved relatively immune from foreign competition and attracted many new converts, especially following the 1933 establishment of the Milk Marketing Board, which stabilized prices. Government officials therefore hoped that in wartime, increased domestic milk production would provide a substitute for foreign meat, butter, and cheese imports.

The wartime demand for milk impacted directly upon Britain's livestock economy. Shortages of imported feed prevented a substantial expansion in the national dairy herd; therefore, improving milk yields was dependent upon a growth in productivity. To this end, the bodily economy of the dairy cow was subjected to enhanced state scrutiny. Applying an industrial model of production, inputs (feed and labor) were weighed against outputs (milk and calves). Cows that fell below the required standards were branded "passengers" and recommended for culling. Encouraged by high set prices and threatened by eviction if they failed to follow official advice, dairy farmers swiftly adapted to this new system.[2]

As a factor that impacted adversely on milk yields, disease was awarded new significance within the context of war. Highly contagious livestock diseases had been targeted by state "stamping out" (or "cull-and-slaughter") policies since the later nineteenth century. During the interwar period, public health concerns over bovine tuberculosis resulted in several new state initiatives. However, diseases of production were traditionally regarded as the farmer's responsibility. Due to a shortage of capital and a lack of regard for veterinary ability, farmers rarely sought professional aid. Instead, ailing cows were marketed, sent to the butcher, or treated with family and patent remedies.[3] To circumvent sterility (the failure to breed) and abortion (the premature termination of pregnancy), many farmers kept "flying herds," maintained by the purchase of freshly calved cows that they sold when milk yields dropped.[4] The increasing frequency with which cows changed hands facilitated the spread of disease. In 1934, the Economic Advisory Council's Committee on Cattle Diseases reported that the average dairy cow survived only half of her useful life. Disease—most importantly reproductive disease—accounted for around half of all disposals from herds and cost farmers 2.5 million pounds a year.[5]

Since cows produce milk only following the birth of a calf, reproductive disease posed an important challenge to the wartime drive for more milk. Moreover, the new focus on productivity meant that interwar responses to disease were no longer appropriate.[6] British veterinarians responded to this situation by devising countermeasures to sterility and abortion and winning state support for their application on farms. Their activities were highly significant: they extended the state's "reach" over wartime agriculture, contributed to the production drive, created new forms of veterinary expertise, and generated new relationships between veterinarians and farmers. Yet, existing historical accounts have largely ignored such developments. Histories of agriculture in wartime tend to regard the transformation of dairy farming as a political affair, the natural outcome of negotiations between government officials and farmers. Consequently, they do not prob-lematize new farming methods or their development and application by new sets of experts.[7] While veterinary histories document the profession's activities, they do not subject them to detailed analysis on the basis that the benefits of veterinary intervention were self-evident.[8] The flawed nature of this assumption is revealed in the following account, which opens with a brief examination of the veterinarian's role in interwar Britain. I reveal that when war broke out, the increased demand for milk did not automatically translate into a demand for veterinary services. Rather veterinary expertise had to be constructed actively and made relevant to the new context. The profession's leaders performed this task through the creation of a "Scheme for the Control of Certain Diseases of Dairy Cows," which gained the back-ing of farmers and the state. Drawing on preexisting but rarely applied tech-nologies, the scheme provided the opportunities and education necessary for new veterinary interventions in cattle breeding. I argue that its operation transformed understandings of fertility, raised veterinarians to the status of experts, and facilitated the shift to a productivity-oriented agriculture.

Veterinary Practice in Interwar Britain

The interwar years proved difficult for many British veterinarians. Tradi-tional modes of employment diminished as the rise of motorized transport led to a decline in horse numbers. The profession's attempts to carve out a new professional niche in meat and milk inspection met with only lim-ited success.[9] Other possibilities included the expansion of agricultural or small-animal practice. The latter provoked an ambivalent response. Veteri-narians often complained about its "sentimental" basis and largely female clientele.[10] Nevertheless, "many of them who would not be seen handling a dog thirty years ago were very glad to see them come round the corner in these times."[11] Agricultural practice was a more traditional activity, en-compassing calving cows and treating those suffering from lameness, acute mastitis, digestive troubles, or milk fever. However, veterinarians faced stiff competition from unqualified animal doctors, the "question-and-answer" pages of agricultural magazines, chemists, patent-medicine vendors, lay "castrators," and state-funded agricultural and veterinary advisors.[12]

After rising steeply during the course of World War I, prices for agricul-tural produce went into decline. A precipitous drop of 34 percent between 1929 and 1933 brought them back to prewar levels. A minority of farmers sought to improve output through judicious feeding, breeding, and disease control. Enlisting veterinary aid, they attempted to rid their herds of costly, endemic diseases, namely, tuberculosis and brucellosis (popularly known as "contagious abortion" on account of its symptoms).[13] However, most farmers responded to the interwar depression by cutting capital and labor costs regardless of the impact on productivity. They had little money to spare for veterinary fees.[14] Veterinarian Mary Brancker (qualified 1937) remembers being summoned by farmers who openly admitted, "I'm not sure if I can pay you." On other occasions, she was paid in eggs or cabbages.[15]

Veterinarians faced an additional problem: many farmers had little faith in their abilities. This view was not unfounded. Until 1934, when the veterinary curriculum was lengthened to five years, teaching focused on the horse, leaving those who entered agricultural practice to learn from their colleagues or from bitter experience.[16] Also, many veterinary remedies were little different from home or patent medicines. On account of these factors, many farmers used veterinarians as a "fire-brigade" service, summoning them only as a last resort, by which time animals were often past the point of recovery. The veterinarian's failure to cure such cases only served to reinforce the farmers' low opinion of professional aid.[17]

Various interwar surveys of livestock health suggested that the diseases of breeding were a major problem. Brucellosis affected 40 percent of herds, and sterility was increasingly prevalent.[18] Such problems were traditionally managed by the farmer, although when prolonged or widespread, the veterinarian's advice was sought. Though veterinary understandings of the diseases of breeding were generally more profound than the farmer's, in practice, their approach was extremely similar. Both used the visible appearance of the vulva and vagina and the presence, absence, or regularity of estrus to assess reproductive status.[19] To correct sterility, they employed vaginal douches, pessaries, and, occasionally, artificial insemination.[20]

During the first three decades of the twentieth century, Swiss, Danish, and American veterinarians made considerable progress in elucidating and correcting sterility in cows. They described how, by inserting a hand into the rectum, the state of the cervix, uterus, and oviduct could be determined. It was then possible to pronounce on the presence and stage of pregnancy, as well as on the cause and curability of sterility. The same technique could be used to remedy pathological conditions: abnormal structures on the ovary that impeded the development of estrus were to be removed by manipulation; uterine infections were tackled by guiding a catheter through the cervix and flushing the uterus with iodine solution.[21]

Use of the new techniques promised to distinguish veterinary abilities from those of the farmer and provide them with an advantage in the crowded marketplace. For this reason, they attracted considerable attention from leading British veterinarians. Men such as Harry Steele-Bodger,

William Miller, George Gould, and J. R. Barker were deeply concerned with the plight of the profession and believed that the new reproductive medicine would "convince them [farmers] that we are not kid-glove veterinary surgeons and that we know our job; it will raise us in their estimation as practical scientists, and last, but not least, but greatest of all, it will advance the glorious science of our profession."[22] However, this group was not representative of the profession as a whole. Whereas most veterinarians experienced an insecure existence, working long hours, single-handedly, in an effort to make ends meet, these individuals taught in the veterinary schools or owned large, well-established agricultural practices. They had the time and money to attend meetings at which foreign experts described the new reproductive medicine, and they managed to persuade wealthy, progressive farmers to employ them on a regular basis to improve herd fertility. They attempted to communicate the latest advances to the remainder of the profession in papers to local veterinary meetings and articles written for the profession's journal, the *Veterinary Record*. However, most practicing veterinarians had neither the facilities to learn the new techniques nor the opportunities to apply them. Consequently, the supervision of cattle breeding remained largely in the farming domain.[23]

Defining a Veterinary Role in Wartime

The interwar crisis in the veterinary profession deepened with the outbreak of World War II. Mass evacuations, air raids, and food rationing resulted in the voluntary euthanasia of many pets. In Greater London alone, four hundred thousand cats and dogs were destroyed in the first four days of the war.[24] One distraught practitioner commented, "My job could have been done as well by the average slaughter man. . . . I can't raise any enthusiasm for telling of the hundreds of beautiful dogs I shot."[25] Another reliable source of part-time income disappeared when the Ministry of Agriculture and Fisheries (MAF) cancelled its "attested-herds" scheme, which had paid practicing veterinarians to perform tuberculin testing on selected herds.[26] Veterinarians could not join the armed forces, as secret fears that Germany would utilize livestock diseases as biological weapons had led the MAF to request the profession's reservation. The rationale for its decision was not made public, leading some veterinarians to assume that the government had plans for their employment.[27] However, while farmers' leaders entered into weekly discussions with the MAF over the expansion of agricultural production, veterinarians waited in vain for instructions.[28] When asked in spring 1940 to consider whether additional disease controls could enhance the milk supply, the government's chief veterinary officer, Daniel

Cabot, concluded that "there is little scope for direct action on the part of the Ministry."[29]

Cabot's attitude was hardly surprising. As already demonstrated, during the interwar period, the activities of the State Veterinary Service and practicing veterinarians were largely unconnected to the pursuit of agricultural productivity. Also, there was no precedent of veterinary participation in wartime agriculture. In World War I, the army's reliance on horse transport had resulted in a heavy demand for veterinarians. By the time the food production campaign began in 1916, the profession was fully employed.[30] In 1940, however, unemployment and underemployment rose steadily to affect a quarter of the profession.[31] Desperate to find a wartime role that would justify their reservation, the profession's representative body, the National Veterinary Medical Association (NVMA) joined with animal-welfare charities to form a National Air Raid Precautions for Animals Committee. However, this organization was refused Home Office support and failed to live up to expectations.[32] Attempts to persuade the Ministry of Food to appoint veterinarians as meat inspectors were also unsuccessful.[33] In his 1940 New Year address to the profession, NVMA President Harry Steele-Bodger lamented, "Never has a new year dawned with less promise for the profession. In no previous conflict has the profession been less capable of seeing what or where its duty lies."[34]

In March 1940, the NVMA appointed a committee to consider how veterinary services could be used to the greatest national advantage. Led by Steele-Bodger, who owned a large midlands practice, it was popularly known as the "survey committee." Members comprised a coterie of like-minded, well-connected, and widely respected veterinarians from the fields of research, education, and practice. Most were enthusiasts of the new reproductive medicine.[35] Inspired by the need to "inaugurate a new charter for the profession," they held frequent, lengthy meetings (Steele-Bodger's son, Alisdair, recollected "my job was to pour coffee into them to keep them awake"). They gathered evidence on the incidence and impact of dairy-cattle disease and surveyed the available control methods. In November 1940, after a two-day and three-night session, their first report was completed. This estimated that over seventeen million pounds or two hundred million gallons of milk were lost each year as a result of four diseases: contagious abortion or brucellosis, sterility, mastitis, and Johne's (a wasting disease). Breeding problems alone were responsible for a loss of eleven million pounds.[36]

The committee argued that, in wartime, the country could ill afford the losses caused by disease and that veterinarians possessed a unique capacity to rectify the situation. It proposed to apply their services under a "scheme for the control of certain diseases of dairy cattle" (popularly known as the "survey scheme.") In exchange for a flat fee, payable by the farmer, practicing veterinarians would attend farms at least four times a year. During visits, they would assess herd health and reproductive status, advise on disease prevention, and perform designated treatments, which included the raft of methods devised by foreign specialists in reproductive veterinary medicine. The MAF would publicize the scheme and provide subsidized brucellosis vaccine. Participating farmers had to keep breeding records and seek professional advice at the first sign of illness.[37]

Far more than a technical document, the survey committee report represented a deliberate attempt to make veterinary expertise relevant to the war effort, by extending the profession's gaze from individual sick animals to the health of the herd. It also aimed to win state patronage for the profession and to restructure farmers' attitudes toward animal health and veterinary intervention.[38] Significantly, by the time it was brought before the MAF and the Agricultural Research Council (ARC, the body that advised on the distribution of government funds for agricultural research) in January 1941, the drive for milk production had entered a new phase. In reducing feed imports beyond anticipated levels, U-boat attacks had impacted adversely on milk yields. Meanwhile, demand had risen on account of the cheap-milk scheme for mothers and infants, introduced in July 1940. Acknowledging the need for enhanced state intervention, the government had raised milk prices in an effort to encourage production. It also commenced feedstuff rationing and ordered a controlled reduction in livestock numbers, on the basis that eliminating unthrifty "passengers" would conserve feed for healthy and productive animals.[39] In publicizing and quantifying the contribution that veterinarians could make to marketable milk supplies, the NVMA scheme was extremely well timed. With its attention now focused upon milk production and its relationship to dairy-cow health, the MAF accepted the case for veterinary intervention and worked to impress this view on the National Farmers' Union (NFU), which was somewhat skeptical of veterinary abilities.[40]

While the NFU and NVMA haggled over fees, representatives of the MAF, the ARC, and the NVMA discussed the technical aspects of the scheme. Brucellosis vaccination was an important sticking point. While vaccines had been available for over twenty years, they were of dubious efficacy. Two recently discovered vaccines reportedly produced much better results. They were the 45/20 vaccine devised by veterinary scientist A. D. McEwen at Wye College of Agriculture and the American S19 vaccine. Neither had undergone extensive testing in the British field. However, under pressure from the ARC, the MAF eventually conceded that their use was justified in wartime. To ensure consistent quality, it took responsibility for production. While McEwen could advise on the manufacture of 45/20, S19 had to be flown in from the United States in a bomber, together with the relevant apparatus and an American veterinary scientist, Dr. Mingle, who instructed staff at the MAF's Veterinary Laboratory.[41]

Although advanced as a remedy to underemployment, the scheme did not gain widespread support within the veterinary profession. Some felt that, as a voluntary measure, it did not go far enough. Others questioned the low fees, the obligations that it placed on them, its restraints on clinical freedom, and the feasibility of working in partnership with lazy, ignorant farmers. Another problem was that the reproductive technologies it prescribed were not well known in the profession. Critics' prime concern was that they would be blamed when the scheme failed to live up to farmers' expectations. Their cautious attitude contrasted with the optimistic claims made by the survey committee, illustrating once more the gulf that separated the elite from the body of the profession.[42]

Survey committee members tried to overcome opposition to the scheme using the NVMA's journal, the *Veterinary Record*. Their considerable influence over the contents of this publication resulted in a succession of editorials extolling the virtues of the scheme. Meanwhile, criticisms were confined to occasional supplements, circulated to NVMA members only. These tactics limited MAF and NFU awareness of dissent within the profession and prevented grassroots veterinary opposition from gaining momentum.[43]

Committee members also sought to secure the success of the scheme by combating the widespread lack of expertise in reproductive medicine. The Development Commission (a government-appointed body, predating the ARC, that distributed funds for agricultural research) agreed to fund the appointment of a cadre of "sterility advisory officers." David Spriggs (qualified 1939) applied "with great enthusiasm, because my salary would go up to £400 a year. Before . . . it was five guineas a week." After preliminary training at the Cambridge University School of Agriculture, he and his colleagues went out into the field to instruct practitioners on the investigation, diagnosis, and treatment of diseases of breeding. They also advised on difficult cases and performed clinical research.[44]

Meanwhile, veterinarians proficient in reproductive medicine toured the country, teaching practitioners and addressing NVMA meetings. They showed a film on sterility produced by the drug company Bayer and presented papers on the etiology, pathology, and therapy of breeding problems that appeared later in the *Veterinary Record*. Gathering veterinarians together in market places and abattoirs and on client's farms, they provided "hands -on" tuition in pregnancy diagnosis and sterility management. Additional demonstrations were provided for farmers, in an effort to advertise both the scheme and veterinary expertise.[45] Steele-Bodger's son, Alisdair, related how

> many a time I would drive father to some market town on cattle market day, and we'd get an audience of local veterinary surgeons if they were keen enough, plus interested farmers, and demonstrate, using cows uteri and God knows what, to show people what the uterus looked like, particularly if there was an abortion. How to pass a catheter and so on.[46]

By 1943, most of the 620 vets who had expressed interest in the subject had received personal tuition.[47] It subsequently became commonplace for those seeking posts through the *Veterinary Record*'s classified section to advertise themselves as "sterility trained."[48]

These events reveal that the introduction of the survey scheme at a time of national crisis accelerated the transfer of reproductive technologies across international boundaries, from the laboratory into the field and from the progressive practices of elite vets to the body of the profession. This process also brought about a highly significant change in terminology. Veterinarians deliberately abandoned the term *sterility* in favor of *infertility* or *temporary infertility*, in an effort to persuade farmers that breeding problems could be cured by veterinary intervention.[49]

The "Survey Scheme" in Operation

In conjunction with other government initiatives such as milk recording and selective culling, the survey scheme reversed the decline in milk yields that had occurred during the first two years of the war.[50] It began to operate in May 1942. Uptake peaked in February 1945, when around seven thousand herds or 10 percent of cattle in England and Wales was enrolled. Over ninety herds were under the care of the Steele-Bodger practice. In Scotland, where farmers had proved more resistant to set fees, just three hundred herds participated in the scheme. However, increasing numbers of cows received attention outside the scheme as farmers invited veterinarians to monitor the reproductive health of their herds. Consequently, the nature of veterinary practice changed even for those veterinarians who had refused to operate the scheme. Formerly, they had rarely intervened in cattle breeding due to the lack of farming demand for their services. Now they had frequent access to herds of cows and numerous opportunities to practice and improve their proficiency in newly learned reproductive techniques.[51]

Paradoxically, given the circumstances under which the scheme was conceived, the major impediment to veterinary intervention in cattle breeding was a shortage of veterinarians. In the spring of 1943, the war opened up the Mediterranean and Far East, where the terrain favored animals over motorized transport. The profession was dereserved, and for the remainder of the war, the army scoured veterinary schools and practices for men of fighting age.[52] Additional MAF initiatives in the field of tuberculosis control and brucellosis vaccination added to the profession's responsibilities and detracted from the prosecution of the survey scheme.[53]

Although formerly unconvinced by the merits of veterinary aid, farmers and their representative bodies readily accepted the scheme.[54] Various factors accounted for their changed outlook. Previously, they had possessed neither the means nor the motivation to seek veterinary assistance. However, in real terms, net farming income trebled over the course of the war. Calling the veterinarian was no longer an unaffordable luxury.[55] At the same time, "farming from Whitehall" introduced a new concept of farming citizenship. In return for set prices and a guaranteed market, farmers were expected to submit to increasing state surveillance and comply with "good" farming practice. The latter encompassed a range of issues, including choice of crops, use of machinery, feedstuffs, fertilizer, and labor and attention to livestock health and productivity. Failure to comply with state directives on these matters could result in eviction.[56] A stream of publicity, issued by the MAF and by agricultural and veterinary experts emphasized to farmers the importance of veterinary advice on livestock breeding. This material highlighted the new reproductive techniques available to veterinary surgeons and the superiority of professional aid over home doctoring. Farmers were also informed of the financial benefits that would accrue from their participation in the scheme and told that the failure to seek veterinary assistance early in the course of illness was "helping Hitler."[57]

The structure and content of the survey scheme were also important in reshaping farmers' attitudes. Its operation brought the NFU and NVMA into closer contact, while encouraging farmers to view their veterinarians as friendly collaborators whose functions extended far beyond the fire-brigade treatment of sick animals.[58] Unlike earlier forms of veterinary intervention, the methods applied for the investigation and correction of sterility were unfamiliar to farmers. Their application revealed that visible events such as estrus behavior or external genital changes were unreliable indicators of reproductive status, as determined by veterinary examination of the genital tract. Moreover, the new techniques proved more effective in curing sterility than did traditional remedies.[59] Consequently, farmers became increasingly willing to delegate control over the reproductive health of their herds.

THIS ACCOUNT has demonstrated how, in World War II veterinarians became indispensable to cattle breeding, won the respect and patronage of farmers and the state, and facilitated the drive for improved milk yields. These outcomes should not be regarded as the inevitable result of the wartime demand for milk. Rather, they depended upon the availability of reproductive technologies, the formulation of a suitable delivery system, the conduct of war, and the personalities and drive of veterinary leaders. Motivated by a desire to find employment for the profession in the greatly changed circumstances of wartime, NVMA leaders had sought to strengthen the connections between veterinary services and livestock productivity. The survey scheme achieved this goal by promoting a new focus upon herd health and reproductive capacity. It also created a new "expert" status for the profession, in propelling the uptake of technologies unfamiliar to the farmer, which transformed understandings of cattle breeding.

At the end of the war, maintaining a healthy and balanced agriculture remained a political priority. A global food shortage threatened, and the need to conserve foreign exchange prevented the resumption of food imports. To encourage farmers to modernize, invest, and improve productivity, Parliament passed the 1947 Agriculture Act, which perpetuated the wartime principle of fixed prices, set on an annual basis after consultation with the NFU.[60] The following year saw the passage of the 1948 Veterinary Surgeons Act, which banned unqualified practice and incorporated veterinary education within the universities.[61] These changes perpetuated the wartime shift toward a productivity-oriented agriculture while recognizing and rewarding the veterinary profession's contribution to this goal.

Enrollments in the survey scheme dropped gradually after the war. Having cleared their herds of disease, many farmers cancelled their contracts. The continuing demands of the MAF and the army left veterinarians overworked and with little time to participate. In addition, the scheme's methods became increasingly outdated as field and laboratory research provided new insights into the causes and correction of sterility. In 1950, the MAF and the NVMA agreed that the scheme should end. That the main critic of this decision was the NFU illustrates the dramatic shift in farming perceptions of veterinary expertise that had occurred over the previous decade. Its spokesman argued that it was vital for veterinarians to continue making regular visits to farms. For the NVMA, however, the scheme had served its purpose. It had transformed the nature of veterinary expertise and the standing of the profession. Having secured a national role and a place on the farm, they no longer needed to court the MAF and the NFU for employment.[62]

Notes

A longer version of this article appeared in *Studies in History and Philosophy of Biological and Biomedical Science* 38, no. 2 (2007): 462–87. I would like to thank John Pickstone and David Edgerton for their comments. I am also indebted to my veterinary interviewees: Mary Brancker, Alisdair Steele-Bodger, and David Spriggs. The Wellcome Trust generously funded the research on which this article is based.

1. John Boyd Orr, *As I Recall* (London: MacGibbon & Kee, 1967); Madeline Mayhew, "The 1930s Nutrition Controversy," *Journal of Contemporary History* 23, no. 3 (1988): 445–64; Francis McKee, "The Popularisation of Milk as a Beverage during the 1930s," in *Nutrition in Britain: Science, Scientists and Politics in the Twentieth Century,* ed. David Smith (London: Routledge, 1997), 50–58.

2. Information in the above two paragraphs was drawn from Ministry of Information, *Land at War—The Official Story of British Farming 1939–44* (London: HMSO, 1945); R. J. Hammond, *Food and Agriculture in Britain, 1939–45* (Stanford, CA: Stanford University Press, 1954); Keith Murray, *The History of the Second World War: Agriculture* (London: HMSO, 1955); Stanley Baker, *Milk to Market* (London: Heinemann 1973); Jonathan Brown, *Agriculture in England* (Manchester: Manchester University Press, 1987); Richard Perren, *Agriculture in Depression, 1870–1940* (Cambridge: Cambridge University Press, 1995); Michael Winter, *Rural Politics* (London: Routledge, 1996); Richard Moore-Colyer, "Farming in Depression: Wales between the Wars, 1919–39," *Agricultural History Review* (hereafter *Ag. Hist. Rev.*) 46, no. 2 (1998): 177–96; John Martin, *The Development of Modern Agriculture: British Farming since 1931* (Basingstoke, England: Macmillan, 2000); and Paul Brassley, "Output and Technical Change in Twentieth-Century British Agriculture," *Agriculture History Review* 48, no. 1 (2000): 60–84.

3. *Animal Health: A Centenary* (London: HMSO, 1965); Michael French and Jim Phillips, "Conflicts of Interests: Milk Regulation, 1875–1938," in *Cheated Not Poisoned: Food Regulation in the United Kingdom, 1875–1938,* ed. Michael French and Jim Phillips (Manchester: Manchester University Press, 2000), 158–84; Anne Hardy, "Professional Advantage and Public Health: British Veterinarians and State Veterinary Services, 1865–1939," *Twentieth Century British History* 14, no. 1 (2003): 1–23; Keir Waddington, "To Stamp Out 'So Terrible a

Malady': Bovine Tuberculosis and Tuberculin Testing in Britain, 1890–1939," *Medical History* 48, no. 1 (2004): 29–48; Abigail Woods, *A Manufactured Plague: The History of Foot and Mouth Disease in Britain* (London: Earthscan, 2004).

4. R. G. Linton, "What Is Good Farming?" *Veterinary Record* (hereafter *VR*) 49, no. 10 (1937): 293–98; "Secrets of a Successful Herdsman," *Dairy Farmer* 10, no. 1 (1938): 19; Practitioner, "Disease in 'Flying' Herds," *Home Farmer* 6, no. 4 (1939): 31; interviews with Mary Brancker, 4 November 2002, and Alisdair Steele-Bodger, 27 January 2003.

5. Economic Advisory Council Committee on Cattle Diseases, "Report," *Parliamentary Papers*, 1933–34, Cmd. 4591, iv, 427.

6. Ideally, cows were expected to produce one calf a year. This included the nine-month gestation period. J. Scott-Watson, *The Cattle Breeders Handbook* (London: Ernest Benn, 1926).

7. In addition to note 2, see P. Self and H. J. Storing, *The State and the Farmer* (Berkeley: University of California Press, 1963); Graham Cox, Phillip Lowe, and Michael Winter, "From State Direction to Self-Regulation: The Historical Development of Corporatism in British Agriculture," *Policy and Politics* 14, no. 4 (1986): 475–90; and Martin Smith, *The Politics of Agricultural Support in Britain* (Aldershot, England: Dartmouth, 1990).

8. *Animal Health*; Iain Pattison, *The British Veterinary Profession, 1791–1848* (London: J. A. Allen, 1983).

9. Hardy, "Professional Advantage and Public Health," 1–23.

10. "Veterinary Surgery as a Career," *VR* 41, no. 127 (1929): 572; Major J. J. Dunlop, "Address," *VR* 48, no. 11 (1936): 3–5.

11. J. Mcqueen, "Comment," *VR* 42, no. 30 (1930): 654.

12. Various autobiographies provide graphic descriptions of veterinary practice prior to World War II: R. Hancock, *Memoirs of a Veterinary Surgeon* (London: Country Book Club, 1954); R. Smythe, *Healers on Horseback* (London: J. A. Allen, 1977); E. Straiton, *A Vet in Charge* (London: J. A. Allen, 1979); and J. Herriot, *All Creatures Great and Small* (London: Pan, 1999). My veterinary interviewees, Alisdair Steele-Bodger, Mary Brancker, and David Spriggs, also provided valuable insights.

13. Tuberculosis, caused by the bacterium *Mycobacterium bovis*, was present in around 40 percent of British herds. The consumption of contaminated meat and milk were viewed as significant threats to the public's health. Brucellosis, caused by the bacterium *Brucella abortus*, had a similar prevalence. It also had public health implications, being the cause of undulant fever in humans, but in comparison to TB, this disease was extremely rare. William Miller, *Black's Veterinary Dictionary*, 1st ed. (London: Waverley, 1927); C. P. Beattie, "Undulant Fever Produced by *Brucella abortus*," *Lancet* 219, no. 5671 (1932): 1002–5.

14. Brown, *Agriculture in England,* chap. 5; Martin, *Development of Modern Agriculture,* chap. 2; Waddington, "To Stamp Out 'So Terrible a Malady.'"

15. Interview with Mary Brancker, 4 November 2002.

16. S. L. Trevor, "Correspondence," *VR* 50, no. 46 (1938): 1578; F. H. Thomas, "Disease Research," *Home Farmer* 6, no. 1 (1939): 15. J. S. Steward's notes on veterinary surgery, taken as a student at the Royal Veterinary College London (RVC) in 1927–28, illustrate the equine bias of the curriculum. RVC archive, Hawkshead campus.

17. Hardy, "Professional Advantage and Public Health," 1–23.

18. John McFadyean, "Veterinary Science," in *Agricultural Research in 1925,* ed. Royal Agricultural Society (London: RAS, 1926), 136; Economic Advisory Council Committee on Cattle Diseases, "Report," *Parliamentary Papers,* 1933–34, Cmd. 4591, iv, 427; W. Miller, "Comment," *VR* 51, no. 36 (1939): 1143–44.

19. The estrus cycle in the cow lasts from twenty to twenty-two days. For one or two days in each cycle, the cow comes into estrus (also known as "heat" or "bulling") and accepts the bull. This period coincides roughly with ovulation and is marked by changes in behavior and the enhanced secretion of vaginal mucus. J. Hammond, *Reproduction in the Cow* (Cambridge: Cambridge University Press, 1927).

20. The question-and-answer pages of the *Dairy Farmer* newspaper, 1929–39, passim, reveal farming and veterinary perceptions of and responses to breeding problems.

21. J. Albrechtsen, *The Sterility of Cows* (Chicago: Alexander Eger, 1911); W. L. Williams, *The Diseases of the Genital Organs of Domestic Animals* (Ithaca, NY: privately printed, 1921); Folmer Nielsen, "Researches Concerning the Aetiology and Pathogenesis of Sterility in Dairy Cows," in *Annual Congress: Papers, Report and Programme,* comp. National Veterinary Medical Association (London: NVMA, 1926), 199–287; D. L. Stewart, "Infertility in Dairy Cattle," *Irish Veterinary Journal* 16, no. 2 (1962): 21–22.

22. L. M. Magee, "Vaginal and Rectal Exploration," *VR* 44, no. 8 (1932): 209.

23. L. P. Pugh, "Irregular Oestrum in the Cow," *VR* 36, no. 7 (1924): 145–49; J. R. Barker, "The Significance of Oestrum in Cattle," *VR* 41, no. 4 (1929): 78–79; J. F. D. Tutt, "Some Remarks on Bovine Sterility," *VR* 43, no. 14 (1931): 451–52; J. F. Craig, "The Treatment of Sterility in the Cow," *VR* 43, no. 28 (1931): 747; J. R. Barker, "Bovine Sterility," *VR* 46, no. 1 (1934): 13–17; W. C. Miller, "Review: Hormones and Pregnancy," *VR* 48, no. 30 (1936): 902–11; interviews with A. Steele-Bodger, 27 January 2003, D. N. Spriggs, 5 January 2005, and M. Brancker, 4 November 2002.

24. "The Unwarranted Destruction of Small Animals," *VR* 51, no. 37 (1939): 1154–55; "Editorial," *VR* 52, no. 28 (1940): 512; A. Moss and E. Kirby, *Animals*

Were There: A Report of the Work of the RSPCA during the War of 1939–45 (London: Hutchinson, 1947), 18; M. Brancker interview, 4 November 2002.

25. Eric Leaver, "Correspondence," *VR* 52, no. 45 (1940): 794.

26. Ministry of Agriculture, "Emergency Notices," *VR* 51, no. 37 (1939): 1155; Waddington, "To Stamp Out 'So Terrible a Malady,'" 29–48.

27. General Policy for Veterinary Surgeons, 1939, LAB 6/132, British National Archives (hereafter NA); National Veterinary Medical Association (hereafter NVMA), "Quarterly meeting," *VR* 51, no. 4 (1939): 124–28; John Grant, "Correspondence," *VR* 51, no. 42 (1939): 1277.

28. National Farmers' Union, *Yearbook* (London: NFU, 1940), 26–27; Correspondence, Nov. 1939, LAB 6/132, NA.

29. D. Cabot memo, March 1940, MAF 52/257, NA.

30. Hardy, "Professional Advantage and Public Health," 7–8.

31. Correspondence, Dec. 1940, MAF 35/592, NA.

32. NVMA, "Quarterly Meeting," *VR* 51, no. 4 (1939): 108 and 124–26; "Editorial," *VR* 52, no. 17 (1940): 325–26.

33. "Editorial," *VR* 51, no. 52 (1939): 1487–88; "Editorial," *VR* 52, no. 16 (1940): 304. Earlier, unsuccessful attempts to gain recognition as meat inspectors are described in Hardy, "Professional Advantage and Public Health," 18–23.

34. Harry Steele-Bodger, "New Year Message," *VR* 52, no. 1 (1940): 1–2.

35. Survey Committee, "Report," *VR* 52, no. 32 (1940), supplement, 41.

36. NVMA, "Report on Diseases of Farm Livestock," *VR* 53, no. 1 (1941): 3–14; H. Steele-Bodger, "Wartime Achievements of the Home Veterinary Services," *VR* 58, no. 50 (1946): 590–91; interviews with A. Steele-Bodger, 27 January 2003, and M. Brancker, 4 November 2002.

37. NVMA, "Report on Diseases of Farm Livestock."

38. H. Steele-Bodger, "Presidential Address," *VR* 52, no. 45 (1940): 797–98; Steele-Bodger, "Wartime Achievements."

39. Murray, *History of the Second World War: Agriculture,* chaps. 4 and 5; Hammond, *Food and Agriculture in Britain,* chaps. 3, 4, and 6; David Smith, "The Rise and Fall of the Scientific Food Committee during the Second World War," in *Food, Science, Policy and Regulation,* ed. David Smith and Jim Phillips (London: Routledge, 2000), 101–16.

40. Meeting reports, 20 Dec. 1940 and 9 Jan. 1941, MAF 35/588, NA.

41. Discussions on the survey scheme, 1940–41, MAF 35/588, NA; Vaccination against brucellosis, 1941, MAF 35/603, NA; "Dr CK Mingle," *VR* 54, no. 20 (1942): 199; Steele-Bodger, "Wartime Achievements."

42. NVMA, "Council and Branch Meeting Reports," *VR* 53 (1941) and 54 (1944), supplements, passim; various, "Correspondence," *VR* 53 (1941) and 54 (1944), supplements, passim.

43. "Editorial," *VR* 53, no. 1 (1941): 14; no. 14, 206–7; no. 35, 511–12; and NVMA, "Council and Branch Meeting Reports."

44. Agricultural advisory work 1940–42, D 4/114 and 4/119, NA; interview with D. N. Spriggs, 5 January 2005.

45. "Refresher Tuition," *VR* 53, no. 44 (1941): 642; NVMA, "Council and Branch Meeting Reports"; Steele-Bodger, "Wartime Achievements."

46. Interview with A. Steele-Bodger, 27 January 2003.

47. In total, there were approximately fifteen hundred active practitioners in Britain. W. Wooldridge, "NVMA Presidential Address," *VR* 54, no. 1 (1942): 1–2.

48. "Appointments Required," *VR* 54, no. 23 (1942): v; "Appointments Required," *VR* no. 17 (1943): v.

49. H. Steele Bodger, "Speech at the Farmers' Club," *Dairy Farmer* 15, no. 12 (1943): 8; listings in *VR* index, 1939–45. Pfeffer notes a concurrent shift in the medical profession's terminology. She argues that rather than branding one partner "sterile," doctors increasingly regarded "infertility" as the product of an incompatible union. This was very different from the veterinary conceptualization of the term. N. Pfeffer, *The Stork and the Syringe* (Cambridge: Polity Press, 1993), 60.

50. Murray, *History of the Second World War: Agriculture,* 176.

51. "Return Showing by Counties the Number of Veterinary Surgeons Participating in the Scheme," *VR* 55, no. 46 (1943): 451; "Editorial," *VR* 56, no. 6 (1944): 46; Survey Committee, "Report," *VR* 56, no. 17 (1944), supplement, 55; NVMA, "Branch Discussions on the Scheme," *VR* 56 (1944), supplements, passim. Veterinary practice records provide evidence of the rising frequency with which veterinarians attended herds to rectify fertility problems. Records of W. B. and T. A. Coe (Bury St. Edmunds Records Office); records of Garston Veterinary Group (in possession of Garston Veterinary Group, Frome, England).

52. F. Bullock, "Correspondence," *VR* 55, no. 23 (1943): 240; The Requirements of the Forces for Veterinary Surgeons, MAF 47/128, NA.

53. "Editorial," *VR* 55, no. 29 (1943): 278–79; F. Chambers, "Correspondence," *VR* 55, no. 37 (1943): 348; "Re-opening of Admission to the Attested Herds Scheme," *VR* 56, no. 27 (1944): 245; "Vaccination of Calves against Contagious Abortion," *VR* 56, no. 44 (1944): 430.

54. Nottinghamshire Farmers Union meeting, 2 April 1942 (archive held at Nottingham County Record Office); Cheshire FU meeting, 9 March 1942 (Cheshire RO); Wiltshire FU meeting, 25 March 1942 (Wiltshire RO).

55. J. K. Bowers, "British Agricultural Policy since the Second World War," *Agricultural History Review* 33, no. 1 (1985): 66.

56. Martin, *Development of Modern Agriculture,* chap. 3

57. "Council of Agriculture Recommend "Panel" Scheme," *VR* 54, no. 33 (1942): 334; J. Mackintosh, "Breeding and Rearing to Maintain the Milk Supply,"

Journal of the Ministry of Agriculture (hereafter *J. Min. Ag.*) 48, no. 3 (1941–42): 151–55; FRCVS, "How to Deal with Sterility," *Dairy Farmer* 14, no. 6 (1942): 15; articles on cattle disease, *Home Farmer* nos. 5–9 (1942), passim; John Hammond and J. Hunter-Smith, "Winter Milk Production," *J. Min. Ag.* 49, no. 1 (1942–43): 3–8; MRCVS, "Points from my Practice," *Dairy Farmer* 15, no. 3 (1943), 15; "The Veterinary Outlook," *Lancet* 243, no. 6283 (1944): 155–56.

58. NFU and NVMA meetings, 1942–50, MAF 35/587, 35/588, and 35/589, NA.

59. "Information," *Dairy Farmer* 14, no. 3 (1942): 19; H. P. Donald, "Heat during Pregnancy in Dairy Cows," *VR* 55, no. 31 (1943): 297–98; J. A. Laing, "Some Factors in the Etiology and Diagnosis of Bovine Infertility," *VR* 57, no. 23 (1945): 275–80; NVMA, "Branch Discussions on the Scheme"; Ministry of Agriculture, "News Service," *VR* 57, no. 29 (1945): 353; replies from county branches, 1945, MAF 35/589, NA.

60. Martin, *Development of Modern Agriculture*, chap. 4.

61. "Veterinary Surgeon's Bill," *VR* 60, no. 18 (1948): 207–12.

62. Reports of meetings, 1947–50, MAF 35/589, NA.; note of meeting, 11 Feb. 1949, MAF 35/58, NA.

CHAPTER 4

Policing Epizootics

Legislation and Administration during Outbreaks
of Cattle Plague in Eighteenth-Century Northern Germany
as Continuous Crisis Management

Dominik Hünniger

EPIZOOTICS, ESPECIALLY of cattle plague, raged through Europe during the eighteenth century with three peaks in incidence. The first occurred from 1711 to 1717, the second from 1745 to 1757, and the third from 1769 to 1786.[1] Their devastating impact on the economy and society can hardly be overrated in the light of estimated mortality rates of between 70 and 90 percent. However, until recently the historiography on veterinary medicine in Germany has concentrated on the development of veterinary services and has been written by practicing veterinarians.[2] Unlike the diverse and original research on livestock diseases in modern Africa,[3] studies on early modern Germany are rare, and hardly any attempt has been made to reveal the effect various diseases have had on the everyday life of rural populations and the enormous challenge epizootics posed to early modern administrations.[4]

This chapter aims to shed new light on some aspects of disease control by focusing on changes and continuities in official legislation in times of cattle plague during the eighteenth century. This disease has become synonymous with rinderpest, but because we do not know exactly what the disease was and retrospective diagnoses are historically suspect, I stick to the contemporary language that described these epizootics as "horned cattle plague" (*Hornvieh-Seuche*). The main body of sources is sixty-eight so-called police ordinances (*Policeyordnungen*) that were published from 1682 to 1798 in the Duchies of Schleswig and Holstein, then under the Danish crown. Similar ordinances can be found in many other territories of the Holy Roman Empire (in fact, in most parts of Europe), and many historians of veterinary medicine have examined these documents. However, the majority of veterinarians who have analyzed these ordinances have assumed that they reflect reliable facts about actual events. Here, I am arguing for a more careful reading of these sources because these ordinances reveal only government intentions as how to deal with an epidemic at a legislative level. In practice, the reactions of local administrations were shaped by a less rigorous approach to disease control, as Jutta Nowosadtko has recently pointed out.[5]

Reflecting upon recent research on police ordinances and state-formation processes, which stresses the contested nature of early modern legislation and administration, I will ask how different groups of actors shaped these ordinances in a communicative process. As a short summary of this research, I want to highlight three aspects that are particularly important for my work. First, administrative action was always concerned with local circumstances and the special needs of dominant social groups. Laws and regulations had to be adjusted to meet the demands of everyday life. Regulations may have been formulated as general and universal, but people assumed that these rules would be open to adjustment in special circumstances and individual cases. Hence, law and practice were part of a circular process; this fact is not so much interesting in terms of the difference between legislative claims and actual (non-) compliance as it is in the way in which different social groups negotiated these regulations and for what reasons.[6]

Second, state formation in the early modern period was not a process of simple, straightforward modernization but rather a continuous one with many ruptures and idiosyncrasies, a process in which authorities and subjects, center and periphery, court and province had to negotiate the extent and limitations of power.[7] Power in early modern times was directed at acceptance; and territorial and local authorities, as well as other corporate bodies and certain individuals, collaborated closely. Generally, all parties involved aimed for consensus but did not avoid conflicts when their livelihoods or interests were at stake.[8]

Finally, early modern political language was a "language of legitimisation"[9] that justified its aims according to generally accepted values. In this respect, although negotiation was almost always at play, it rarely happened between equals, and power was, of course, distributed unevenly. However, individuals were always involved in acts of persuasion when they wanted to exercise political power. According to Michael Braddick, "These acts of persuasion are best observed in micro-historical contexts—the face-to-face situations in which claims to political power are actually asserted and tested. . . . In face-to-face situations the claim to political power is not usually imposed by force, but is negotiated."[10]

In accordance with these lines of research, early modern society's reactions to epizootics can provide insights into processes of state formation because epizootics and their containment severely disrupted everyday life. The enforcement of ordinances for the control of animal epidemics, like their predecessors for bubonic plague in humans, involved an interaction between the interests of merchants, artisans, magistrates, rulers, and local authorities.[11] I attempt to show how the early modern state "managed" epidemics by applying legal measures and how various groups and individuals contested them. My focus will be on the rhetoric of the decrees as well as on their content. By looking at general ordinances and other legislation that dealt with specific problems, I explain how animal-disease regulations represented a continuous form of crisis management. Some of the more specific decrees were issued after older and more general ordinances had encountered obstacles either because of opposition from the subjects themselves or because they threatened to undermine a flourishing economy. For these reasons, quite a few ordinances had to be revised as individual epizootics progressed. The evidence suggests that ordinances resulted from two types of situations. First, a decree that covered general aspects of disease control was published once the authorities had discovered an outbreak of cattle plague. Second, once a piece of legislation proved either unworkable in practice or faced considerable opposition, the authorities revised it.

Decrees for cattle-plague control sometimes began with statements about the origin and course of the disease in neighboring countries or parts of the legislators' territories; their purpose was to prevent the further spread of the disease. To justify strict edicts, authorities would elaborate on their responsibilities for the well-being of their subjects and territory by referring to their "sovereign precaution" (*Landes-Väterliche Vorsorge*). This terminology can be found in almost every decree and is the most widely used term in the early modern "language of legitimisation." Most of the more general decrees were very detailed and have paragraphs on various aspects of plague control, such as preventive slaughter, drugs, cleaning and hygienic measures, and methods of diagnosis. I am not going to dwell on the wide range of measures in detail, but rather concentrate on certain key aspects that were deeply contested, namely, trade regulations, quarantine measures, and the use of animal products.

The Regulation of the Livestock Trade

Since the Middle Ages, the cattle trade between Denmark and the Netherlands had been a major contributor to the economy of the Duchies of Schleswig and Holstein.[12] The ox trade accounted for 70 to 80 percent of the yearly revenues at the duchies' main toll station at Gottorf in Schleswig. In Denmark, farmers raised oxen on pastures for four to five years. During the following winter, these animals were fattened on noble and royal farms. Then, primarily Dutch cattle dealers purchased them and moved them south to the Netherlands, via the duchies. At times, up to fifty thousand head of cattle per year traveled along the so-called oxen routes (*Ochsenwege*) that ran through the duchies on the central lowlands between the marshes in the west and the hills in the east. This trade peaked in the seventeenth century and gradually decreased at the end of the eighteenth due to the protectionist economic policies of the Dutch and Danish authorities and, more importantly, because serious outbreaks of cattle plague encouraged a stronger reliance on homebred cattle. Economic protectionism was thus linked to measures of disease control and showed that eighteenth-century authorities drew their conclusions from the enormous losses during cattle-plague outbreaks.

Given the importance of the cattle trade, many of the disease-containment regulations specifically concerned trade and markets. The issuing of health certificates for any cattle that moved across the land was probably the most important measure concerning trade in the eighteenth century. This regulation first appeared in the oldest surviving police ordinance that specifically related to the cattle plague. On 1 November 1682, the duke of Schleswig and Holstein, Christian Albrecht, issued an ordinance that was to be the model for every new decree during the eighteenth century. The duke forbade the movement of livestock to pastures and between stables without a health certificate. Another aspect of these regulations was the ban on cattle imports when news of an epidemic in nearby countries reached Schleswig-Holstein, as was the case in the period 1729–31 when the cattle plague hit Poland and the neighboring territories of Brandenburg and Mecklenburg.[13] The ordinances issued between 1729 and 1731 set high penalties for anyone who transgressed a rule and explicitly ordered local authorities to comply rigorously with the legislation. A ban was also introduced restricting the trade in animal products (like meat, skin, and hair), commodities that could not be imported unless the owners had certificates confirming the animals' good health while alive. To encourage obedience, the authorities ordered that any cattle that reached the duchies without a health certificate had to be slaughtered and buried immediately, so no profits accrued to the trader. However, although the authorities considered trade restrictions to be efficient methods of disease control, they were aware of the need to keep the economy going and tried to achieve this by guarding and quarantining only infected villages, rather than the whole duchy. The authorities informed ox traders about infected areas and prescribed roads through unaffected places so that cattle movement from Jutland and Denmark would be "safe and without any danger."[14]

Similar measures were introduced at the start of the most severe outbreak of cattle plague in the eighteenth century, the epidemic of 1745–46. The complete cancellation of cattle markets was the topic of just one of twenty paragraphs of the general ordinance of 5 February 1745.[15] However, the cattle trade was not banned entirely but could continue with all the necessary precautions, such as traders' having health certificates and avoiding roads that ran through infected places.[16] An ordinance of 19 February 1745 showed that the authorities planned to limit the movement of humans as well as cattle because they assumed that people were also carriers of this dreaded disease. The first paragraph stated that anyone wishing to travel had to have a certificate with details about his or her state of health, stature, age, color (of hair, face, and eyes), facial features, and clothes. Regarding animals, their color and number had to be written on the certificate, and cattle owners and drovers had to confirm by oath that the herd "neither has any signs of the disease, nor was taken from a byre, house or neighborhood, where the epidemic was felt in some way or another" and "he [who trades the animals] himself has not been at a suspicious place."[17] The government placed much of the onus for enforcing the rules on the toll keepers who were expected to check the certificates and the goods that passed through their gates.

On paper, the punishments for infringements were strict: those who possessed counterfeited certificates were to be prosecuted and sentenced to lifelong hard labor, while public servants who had issued illegal or unsound certificates were fined.[18] The respective severity of the legislation suggests that responsibility lay with the cattle trader rather than with the civic authorities. Additionally, the ordinance of 8 December 1746, which dealt with illicit trade, explicitly harked back to the regulations of the 5 and 19 February ordinances, as well as to that of 11 October 1745. The rhetoric of this ordinance was very vivid and evoked a colorful picture of the smugglers:

"Mischievous and avaricious cattle dealers from Jutland had deliberately and because of cheap prices traded with suspicious livestock and through their greed had propagated the disease in a spiteful manner."[19] This decree illustrates how the language of these ordinances became more drastic the longer it took the authorities to bring the epizootic under control.

Little changed as the eighteenth century progressed. The cattle plague that had started in February 1745 raged through much of the duchies for at least one year and lingered in some provinces until the spring of 1752—seven years after the first cases. The end of this epizootic was marked by the ordinance of 6 May 1752, which stated that the disease had decreased and the authorities tried to galvanize trade by authorizing a return to normal commerce.[20] Nevertheless, in an effort to prevent further introductions of the disease or another flare-up, the authorities still demanded health certificates for the movement of cattle, showing how legislation introduced during an epizootic had legal repercussions that went far beyond the cessation of an outbreak.

Unfortunately, it was only ten years before the authorities had to reintroduce stricter legislation to deal with a further outbreak. On 12 October 1762, all cattle markets were prohibited because cattle plague had appeared in the southern parts of Denmark, bordering the Duchy of Schleswig. This time the epidemic seemed to have raged for a shorter period, and mortality was not as heavy as in the outbreak of 1745–46. Then, an epidemic similar in severity to that of 1745–46 broke out in March 1776 and persisted in the duchies for ten years, peaking in 1779. This time, the authorities issued a twenty-page "General and Constant Ordinance against the Horned Cattle Plague" (*Allgemeine und Beständige Anordnung gegen die Hornvieh-Seuche*) on 7 March 1776. This decree was similar to laws introduced thirty years earlier. Concerning disease-control measures, there was hardly anything new, but what was striking was the greater amount of detail devoted to the regulations and the greater emphasis on penalties, reflecting a rhetorical desire on the part of the authorities to be more rigorous than ever.[21] In summary, the length of time it took to eliminate the cattle plague indicated how difficult it was for countries in the early modern period to tackle disease, not least because authorities and affected people alike had to consider economic, social, and political circumstances and sensibilities.

Quarantine

Besides trade regulations, decrees dealing with the segregation of healthy animals outside the villages had the most drastic impact on society and the economy. Quarantine measures can already be found in the 1682 decree.

From 1718 onward, the general ordinances often contained the following types of regulation. First, apparently healthy animals had to be separated from sick bovines and placed in special huts located outside the villages.[22] Next, authorities appointed guards to supervise the cattle; these men were also separated from the village community, living in huts on the grazing lands outside the villages. In addition, if a farm was infected, it was separated from the rest of the community by special quarantine regulations, which mirrored what happened during outbreaks of the bubonic plague. Guards had to watch infected houses and ensure that nobody entered or left the premises. Neighbors were expected to supply food and other provisions, which they deposited at a place some distance from the house. If the disease reached two or three of the bigger farms, the whole village was shut off and had to be guarded by neighboring villagers if their village was not yet infected. Ordinary subjects who were employed as sentinels and guards for infected villages played a major role in disease-containment action. However, their unreliability was apparent almost from the start. To try to keep these subjects loyal, the decree of 1 March 1745, section 2, ordered that sentinels and guards had to swear special oaths to the authorities, stating that they would abide by the rules. Yet, because this decree was widely disregarded, authorities increasingly called upon the militia to enforce the quarantine in the course of the spring and summer of 1745. Later, however, when the plague had still not ceased in October 1745, the government officially disbanded all sentinels and guards.[23]

The quarantining of whole villages from the rest of the country caused serious problems as can be seen in the letters and supplications of farmers from the parish of Tetenbüll in the district of Eiderstedt from 1745 to 1746.[24] These letters provide detailed insight into everyday lives during times of crisis. An example of these tensions is apparent in the petition of a cobbler from Tetenbüll, Dirk Asmus. Asmus's work depended on the use of cattle hides. Complaining of the economic difficulties that arose because he could no longer earn money from trade, he asked for a concession to allow him to send already-manufactured boots and shoes out of the quarantined village. Another major problem was that an infected village was almost under a state of siege, so there were severe food shortages (especially flour). Food was not distributed equally, and many peasants complained that the guards took more than their fair share of provisions, calling them "impertinent eaters and drinkers"[25] because they demanded bacon and meat instead of porridge. Psychological hardship affected members of small religious denominations, such as the Mennonites and Remonstrants, who were unable to attend the usual services at Friedrichstadt about fifteen miles away. Instead, they had to take a special oath to enable them to attend the Lutheran church in the village. These are just some of the effects quarantine had on those who had to endure the sequestering of their villages. Similar complaints, as well as petitions demanding changes to the ordinances, reached the authorities from every corner of the duchies, and these increased in number and frequency the longer an epizootic prevailed.

The ordinance of 19 February 1745 is an example of how legal documents can be a good source of information on problems surrounding the implementation of these decrees. Its preamble stated that new developments and reports from the provinces require the extension or restriction and general modification of the general decree of 5 February 1745. The authorities realized that, in practice, many of the laws were unworkable; and for their own credibility, as well as to appease members of the community, they adjusted measures that were impossible to enforce or were abused. This is also clear from the decree of 19 January 1779, which listed a number of exceptions to the complete sequestering of villages. Help from outside was allowed in cases of fire. Priests and doctors could visit a quarantined place when needed in cases of illnesses and deaths, and midwives could attend expectant mothers, all, of course, with the necessary precautions.

The Ban on Skinning

Animal products like milk, skin, hair, horns, and manure played an important part in the rural economy of eighteenth-century Schleswig and Holstein. Oxen were traded mainly for their meat, but the animal products mentioned above also played a vital role in the local economies. In addition, farmers used oxen for drafting, so quarantine measures had an important impact on trade as well as on plowing. Thus, in an economy that was based on the utilization of every part of the animal, the policies of culling not only met with resistance from livestock owners but also caused serious economic problems for the duchies.

Throughout the eighteenth century, the more general decrees forbade the skinning of infected cattle and ordered all carcasses to be buried completely and be covered with lime. However, the authorities often repealed or modified these prohibitions due to the shortage, and thus high costs, of leather and leather goods. This attitude showed how the authorities tried to counterbalance two important issues: plague control and the promotion of economic stability and wealth. These two issues had to be negotiated between different parties time and time again.

In the eighteenth century, because of popular concepts of illness and contagion, dead bovines were also seen as potential carriers of the cattle plague. So, when the disease broke out in February 1745, the first paragraph of the first ordinance demanded the complete burial of any infected animal's carcass, four to six feet deep in the ground and covered in unslaked lime as a disinfectant. The authorities also deemed hides to be possible carriers of the disease and introduced a ban on their importation into the Duchy of Holstein on 10 February 1745. Strict adherence to the decrees pushed subjects as well as the authorities to their economic limits, and there was a severe shortage of hides. To try to protect the local economy, the Duchy of Holstein issued a new decree on 6 July 1745, which prohibited the exports of hides and leather from the duchies to ensure that local communities had access to these products.

It was not only hides that became scarce, so too did the fat of animals, used to make tallow for candles. On 14 December 1745, the Danish king issued a special decree for the Duchy of Schleswig on the usage of tallow.[26] Inhabitants were now permitted to take the tallow from dead animals if they did so in remote places and out in the open. This new measure was introduced not only to respond to popular petitions but also to steady prices by putting an end to the speculation in animal products. This measure followed earlier edicts aimed at curbing speculation in foodstuffs, which had led to high prices in the towns. In August 1745, two decrees had forbidden the preemptive buying of large quantities of butter from farmers to sell in small quantities in the cities at inflated prices. In the early modern period, ensuring people had access to affordable food was necessary to uphold public order and forestall revolts that could threaten the stability of the state.

The decree of 27 April 1764 was especially interesting as it provided details about how skinners were to prepare the carcasses. The skin was to be covered with lime or ash and had to dry in the open air away from other animals Hides could not be sold or moved for a six-week period. The raw tallow had to be melted at the place of removal and prepared before it could be taken home. Further regulations appeared in response to the serious epidemic that broke out in 1776. In a general decree of 7 March 1776, as many as four paragraphs and two supplements were dedicated to regulations about skinning and the use of animal products. These included more details on the procedures for the removal of fat and skins. The skin was to be removed at the burial site and near a watercourse, within twenty-four hours after the animal died. Tails, ears, horns, and hooves had to be buried with the carcass. The removed skins had to be washed and put into limewater for fourteen days until the hair fell off. This hair then had to be buried as well. If the skin could not be processed and tanned immediately, it had to be put in a remote place for drying. Exportation was allowed only after fourteen days. People dealing with the carcasses were ordered to wear special oilcloths. The whole process was to be watched over by two inspectors; everyone else was restricted: "dispensable people who are only spectators are not admitted."[27]

Whether these regulations were enforceable in practice, they were nonetheless onerous in terms of time and labor and demonstrated how disease-control measures involved not only restriction on trade and quarantines but had a significant impact on the working lives of rural communities and indeed the territory as a whole. When petitioners asked for the modification of certain decrees, their arguments were mainly economic and fiscal. If the king did not allow certain works to continue, his subjects argued, they would be unable to earn money and therefore incapable of paying their taxes. Not surprisingly, early modern princes often submitted to this line of reasoning.

The Rhetoric and Practice of Livestock Regulations

By evaluating the regulations on cattle epidemics, I have illustrated how an early modern government attempted to control disease through legislation. Different interests on the part of authorities and subjects had to be negotiated and found their expression in either stricter or more detailed ordinances. Hence, animal-disease control in early modern Europe can be understood as a complicated interactive process of continuous crisis management in which legislation was altered in response to the practicalities of enforcing it on the ground. Without oversimplifying, I would argue that, at the beginning of an epidemic, the initial laws prohibited activities that the authorities considered to be possible agents for the spread of the disease. However, in the course of time, once an epidemic had prevailed for several months, these strict measures had to be adapted to accommodate other interests as well. When a serious epizootic continued for several years, as in the case of the 1745–52 outbreak, the authorities eventually surrendered to the fact that efforts to contain the cattle plague had failed, as it had spread throughout the duchies. Again, epizootics happened in a social, economic, and political environment where all parties involved and affected not only had to tackle the immediate impact of the diseases but were eventually forced to contextualize disease within a wider variety of circumstances. Subjects expected authorities to be concerned for their well-being, and administrators sought popular approval whenever possible. The rhetoric of the decree of 20 September 1745 is a perfect example of this: "After the harmful cattle plague had spread in this territory so strongly and this contagion, alas!, became almost overarching, and therefore the ordered precautions for containment cannot be of any value anymore, We will change proceedings in order to disburden Our beloved and faithful subjects in these calamitous times."[28]

Forty years later, the official language seems to be very different at first glance. The ordinance of 25 August 1786 that marked the end of the 1776–86 epizootic, suspended earlier prohibitions and stated that "the cattle plague, which broke out years ago in the Duchies, had been checked by the legislative measures."[29] Nonetheless, in practice, although the disease had disappeared by 1786, the regulations had been difficult to uphold, just as they had been during earlier outbreaks. In 1779, a military cordon between Jutland and the Duchy of Schleswig had been established, guarded by six hundred infantrymen and two hussar squadrons.[30] However, that did not prevent people from breaking through the cordon as can be seen in an ordinance dated 23 February 1779, which allowed sentinels to use force and weapons, albeit with the greatest caution and only in extreme emergency. According to the decree, if anyone was injured or died, it was his or her own fault.[31] Certainly, the language used in the 1770s and 1780s can be considered more drastic and self-confident. The preamble of the very first decree at the onslaught of the new outbreak from 7 March 1776 explicitly makes the claim that the regulations were durable, adequate, and "fair to every occurring situation." Nevertheless, over the next ten years, the Danish king had to issue seven modified decrees, and the exceptions, alterations, and sanctions for individuals and collective bodies altered frequently over the years.

Three different aspects of disease control, trade regulations, quarantine, and the restrictions on skinning provide important examples of the interrelationship between law and practice in eighteenth-century Schleswig-Holstein. Each of these issues sheds a different light on the problems surrounding epizootic containment. Since the authorities conceptualized the disease as contagious (even though the nature of the somewhat mysterious contagion remained unnoticed), all contact between healthy and sick animals had to be prevented, and trade was often banned completely at the first sign of disease in the duchies or in neighboring territories. However, as the epidemic continued unabated, these measures were relaxed to keep the economy working. One measure that facilitated trade, while trying to prevent the spread of disease, was the introduction of health certificates. The same applied to the regulations concerning animal hides and products. Instead of an outright ban on the trade in hides and tallow, eighteenth-century decrees tried to make the skinning process safer by introducing measures intended to eliminate a source of contagion.

It is tempting to see these developments as a kind of conscious learning process on the governments' part, as they responded to economic and social realities. But the quarantine regulations attracted considerable resistance from subjects who petitioned the authorities to moderate the rules. Popular complaints exemplified a variety of problems, like shortages of food and animal products. Although the authorities considered a complete quarantine of farms and villages to be the most effective means of containing disease, they were unable to shut off whole villages for a long time. In addition, many of the regulations were subject to infringement and abuse. The repetition and extension of certain laws were evidence of this. At the same time, these extensions can also be understood as the authorities' attempt to prove that they had the capacity to act and control the disease and can be regarded as indicative of the evolution of state building in early modern Europe. Regulations for epizootics clearly show that this process was a complicated combination of bottom-up and top-down developments. The evidence further illustrates that a dreadful crisis like an outbreak of cattle plague shook the foundations of eighteenth-century society and that governments addressed this by mobilizing economic as well as social, cultural, and legal resources. Only by looking at all four aspects is one able to understand the impact of epizootics on agrarian societies.

Notes

This chapter presents the first results of a PhD project on the history of everyday life and governmental action during times of cattle plague in eighteenth-century Schleswig and Holstein. The dissertation was submitted in October 2008 and defended in March 2009 at the University of Göttingen. To date, there has been no study with a similar focus. The veterinary dissertation by Almuth Wagner, "Die Entwicklung des Veterinärwesens im Gebiet des heutigen Schleswig-Holsteins im 18. und 19. Jahrhundert mit besonderer Berücksichtigung der Tierseuchen" (VMD diss. Freie Universität Berlin, 1992), is mainly concerned with the development of veterinary services and the profession. I would like to thank Karen Brown for her invaluable help in improving and clarifying this paper. All remaining shortcomings are, of course, my own.

1. For a detailed, but at times rather ahistorical, overview on the history of cattle plague in early modern Europe, see Wilhelm Dieckerhoff, *Geschichte der Rinderpest und ihrer Literatur: Beitrag zur Geschichte der vergleichenden Pathologie* (Berlin: Enslin, 1890); and Clive A. Spinage, *Cattle Plague: A History* (New York: Kluwer, 2003). Years of outbreaks in the different territories of Europe varied, of course.

2. See the most recent German textbook on the history of veterinary medicine: Angela von den Driesch and Joris Peters, *Geschichte der Tiermedizin:*

5000 Jahre Tierheilkunde (Stuttgart: Schattauer, 2003). Studies with a similar focus as my own research can be found in the studies by Kai Hünemörder, Jutta Nowosadtko, and Karl Peder Pedersen in *Beten, Sammeln, Impfen: Zur Schädlings- und Viehseuchenbekämpfung in der Frühen Neuzeit,* ed. Katharina Engelken, Dominik Hünniger, and Steffi Windelen (Göttingen: Universitätsverlag 2007).

3. In addition to studies in this volume, see especially Richard Waller, "'Clean' and 'Dirty': Cattle Disease and Control Policy in Colonial Kenya, 1900–40," *Journal of African History* 45, no. 1 (2004): 45–80.

4. Only Peter Albrecht, *Die Förderung des Landesausbaues im Herzogtum Braunschweig-Wolfenbüttel im Spiegel der Verwaltungsakten des 18. Jahrhunderts, 1671–1806* (Braunschweig: Waisenhaus, 1980), provides one chapter on governmental action toward epizootics in Braunschweig-Wolfenbüttel (a territory south of the area I am focusing on).

5. Jutta Nowosadtko, "Die policierte Fauna in Theorie und Praxis: Frühneuzeitliche Tierhaltung, Seuchen- und Schädlingsbekämpfung im Spiegel der Policeyvorschriften," in *Policey und frühneuzeitliche Gesellschaft,* ed. Karl Härter (Frankfurt/Main: Klostermann, 2000), 297–340, esp. 318.

6. With a special focus on criminal law: Karl Härter, "Social Control and the Enforcement of Police-Ordinances in Early Modern Criminal Procedure," in *Institutionen, Instrumente und Akteure sozialer Kontrolle und Disziplinierung im frühneuzeitlichen Europa—Institutions, Instruments and Agents of Social Control and Discipline in Early Modern Europe,* ed. Heinz Schilling (Frankfurt/Main: Klostermann, 1999), 39–63; and in German, the more general and very instructive papers by André Holenstein, "Die Umstände der Normen— die Normen der Umstände: Policeyordnungen im kommunikativen Handeln von Verwaltung und lokaler Gesellschaft im Ancien Régime"; and Achim Landwehr, "Policey vor Ort: Die Implementation von Policeyordnungen in der ländlichen Gesellschaft der Frühen Neuzeit," both in Härter, *Policey und frühneuzeitliche Gesellschaft,* 1–46 and 47–70, respectively.

7. Dagmar Freist, "Einleitung: Staatsbildung, lokale Herrschaftsprozesse und kultureller Wandel in der Frühen Neuzeit," in *Staatsbildung als kultureller Prozess: Strukturwandel und Legitimation von Herrschaft in der Frühen Neuzeit,* ed. Ronald G. Asch and Dagmar Freist (Köln: Böhlau, 2005), 1–46, here 13.

8. Stefan Brakensiek, "Lokale Amtsträger in deutschen Territorien der Frühen Neuzeit: Institutionelle Grundlagen, akzeptanzorientierte Herrschaftspraxis und obrigkeitliche Identität," in Asch and Freist, *Staatsbildung als kultureller Prozess,* 49–67.

9. Michael Braddick, "State Formation and Political Culture in Elizabethan and Stuart England: Micro-Histories and Macro-Historical Change" in Asch

and Freist, *Staatsbildung als kultureller Prozess,* 69–90. Although Braddick analyzes the role of local magistrates in the state-formation process, his remarks
about early modern political "language as legitimisation" (69) are very useful
in the context of this paper as well.

10. Ibid., 70.

11. Nowosadtko, "Die policierte Fauna," 317. Jutta Nowosadtko argues that
police ordinances against epizootics, well into the eighteenth century, combined a whole range of measures because of uncertainties about the nature
and cause of epidemics—similar to their model: ordinances against bubonic
plague (322). Although I agree with many aspects of Nowosadtko's research,
my focus is different insofar as I concentrate on the impact of epidemics on the
management of everyday life and Nowosadtko elaborates on questions about
animal husbandry and the human-animal relationship in general.

12. For extensive histories of cattle trade in early modern northwestern Europe, see Wilhelmina Maria Gijsbers, *Kapitale ossen: De internationale handel
in slachtvee in Nordwest-Europa, 1300–1750* (Hilversum, Netherlands: Verloren,
1999); and Johann Bölts and Heinz Wiese, *Rinderhandel und Rinderhaltung im
nordwesteuropäischen Küstengebiet vom 15. bis zum 19. Jahrhundert* (Stuttgart:
Fischer, 1966).

13. Cf. the decrees of 21 December 1729, 11 December 1730, and 3 April 1731.
All ordinances quoted here are held in the manuscript collection of Schleswig-
Holsteinische Landesbibliothek in Kiel. All translations from the German
originals are my own.

14. Decree of 21 March 1718.

15. The extension of 19 February 1745, also forbade horse and small-animal
markets.

16. Additionally, this decree prescribed the eviction of beggars and the foreign poor. I just want to note here that on 20 March 1745, the general ordinance
on begging of 7 September 1732 was renewed and that the time of its renewal
had not been chosen by chance but had been certainly influenced by the raging
cattle plague.

17. Decree of 27 February 1745, §1.

18. Ordinance of 11 October 1745, §§7–10. Later, in 1749, the toll keeper of
Friedrichstadt, Gerhard Oldenburg, was severely punished because he was
involved in illegal trading of infected cattle. "Deductio tam Facti quam Juris
pro fundanda Actione publica wider Gerhard Oldenburg: Controlleur bey der
Königlichen Zoll-Stäte zu Friedrichstadt," in *Puncto Contraventionis des, wegen
der Vieh-Seuche, und des verbothenen Schleich-Handels, abgegebenen allerhöch-
sten Edicti, vom 8ten Decembris 1746. und daher verwirkter Strafe* (Hamburg:
n.p., 1749).

19. Ordinance of 8 December 1746.
20. Ordinance of 6 May 1752. Another decree of 17 March 1779 makes this even more explicit. It refers to the importance of cattle trade for the marsh districts especially and the interest of the authorities in balancing plague control and economic development. The decree then prescribed health certificates and close surveillance by officials.
21. Notably novel was the stipulation forbidding the export of livestock that had survived the last epidemic. This must mean that the authorities believed these animals had acquired some immunity to the disease and that they were thus very valuable for the economy and breeding purposes. This decree even encouraged livestock owners to take their immune cattle to provinces where the epizootic prevailed. The measure can be seen as a precursor to official efforts in the nineteenth century to encourage the breeding of fitter and more productive livestock. In a commentary on the ordinance of 3 January 1763, Eduard Ambrosius, the editor of a collection of Schleswig-Holstein ordinances from 1579 to around 1800, considered the culling of animals unnecessary and even counterproductive because it inhibited infection. Eduard Ambrosius, *Chronologisches Verzeichniß über verschiedene Königliche und Fürstliche Verordnungen und Verfügungen für die Herzogthümer Schleswig und Holstein von 1579 an bis 1792: Mit einem kurzen Inhalt derselben, einigen Anmerkungen und einem Sach-Register, Teil 5: Von 1760 bis 1770* (Schleswig: Serringhausen, 1801), 25. For contemporary discourses on breeding in England, see Nicholas Russel, *Like Engend'ring Like: Heredity and Animal Breeding in Early Modern England* (Cambridge: Cambridge University Press, 1986).
22. The decree of 21 March 1718 banned farmers from moving their cattle on to communal grazing ground and from taking them to infected villages and stalls. This decree also included two interesting paragraphs concerning the assessment of an animal's or village's state of health. In order to find out whether farms were infected, susceptible cattle from disease-free places were brought to these properties to see if "something contagious will show itself." However, the authorities omitted this "experiment" from later decrees, presumably because it was too risky, potentially causing the unnecessary loss of valuable livestock.
23. Ordinance of 11 October 1745.
24. These letters are held in the Kreisarchiv für den Kreis Nordfriesland in Husum, KANF A 2 Ksp. Tetenbüll 95, and a selection has been published by Johan Redlef Volquardsen, "Die Viehseuche in Eiderstedt, besonders im Kirchspiel Tetenbüll im 18. Jahrhundert," in *Nordfriesisches Jahrbuch* 2 (1966): 101–09. For a Danish peasant's perspective on the cattle plague, see Karl Peder Pedersen, "Cattle Plague and Rural Economy in 18th Century Funen: The Notebook of the Funen Tenant Peder Madsen at Munkgaarde: Applications and

Perspectives," in *Bäurliche (An-)Schreibebücher als Quellen zur Wirtschaftsge-schichte,* ed. Klaus-Joachim Lorenzen-Schmidt and Bjørn Poulsen (Neumün-ster, Germany: Wachholtz, 1992), 73–88.

25. KANF A 2 Ksp. Tetenbüell 95.

26. A similar decree for the Duchy of Holstein was published on 20 January 1746.

27. Decree of 7 March 1776, supplement C.

28. Decree of 20 September 1745.

29. Ordinance of 25 August 1786.

30. Decree of 5 January 1779.

31. Decree of 23 February 1779.

CHAPTER 5

For Better or Worse?

The Impact of the Veterinarian Service on the Development of the Agricultural Society in Java (Indonesia) in the Nineteenth Century

Martine Barwegen

MANY AUTHORS researching the agricultural history of Java or Indonesia mention the importance of livestock and its contribution to economic developments. However, they do not provide a detailed study of the historical, economic, and ecological significance of the livestock sector. In this chapter, I will explore the impact of the veterinarian service and its contributions, both positive and negative, to combating cattle plague.

The sporadic occurrence of cattle plagues (rinderpest) in the seventeenth century has been overshadowed by the great epidemics of human plague. Veterinary medicine as such was nonexistent at that time, both in theory and practice apart from a few treatises on diseases of horses, reflecting the preoccupation of the ruling classes in Europe with this important transport animal. In the nineteenth century, this was still the case in Indonesia. In 1814, the first military horse veterinarian arrived in Java. The corps of military horse veterinarians grew to five men in 1851, just before the establishment of the civil veterinary service (Burgerlijke Veeartsenijkundige Dienst [BVD]) in 1853. These military horse veterinarians had hardly any interest in the livestock belonging to Javanese farmers, although attending to their animals was part of their job.

With the establishment of the civil veterinary service, Javanese livestock holders could utilize a form of science that was new to their culture. Javanese cattle holders had always made use of knowledge inherited from their ancestors and based on their own experiences. Lise Wilkinson, with just one sentence, inspired me to look further into the question as to whether the development of the veterinary service in the nineteenth century actually benefited Indonesian cattle owners, especially when the cattle plague raged over Java in 1878. She wrote, "It is also possible that climatic factors and political difficulties added to the seriousness of the situation and to the complexities faced by the authorities."[1] Unfortunately, that was all she wrote because it was never questioned whether the status and condition of the livestock improved after the introduction of the veterinary service.[2] On the other hand, the impact of military horse veterinarians was questioned by Groeneveld in 1916:

> Isn't it peculiar that everywhere a breed breeds itself, it has so many good characteristics, and isn't it typical that greatest degeneration is to be found in the areas where Europeans tried to improve the breeds and where castration was practiced? . . . Then we have to acknowledge that it is exactly through our interference—direct and indirect—that the present situation is created and that nowhere the dawn of the day glories.[3]

In the context of the civil livestock economy, the question arises as to why the outbreak of the cattle plague in 1878 ruined the West-Javanese agriculture and led to famine. What was the effect of official policies aimed at suppressing this epizootic? In this chapter, the contribution of the veterinary service to the development of the Javanese agricultural society will be studied using the cattle plague as an example. To answer these questions, first an overview of the development of the veterinary service will be given, followed by a description of the outbreak of the cattle plague in 1878.

The Establishment of the Civil Veterinary Service (BVD)

In Europe, veterinary schools appeared at the end of the eighteenth century. One of the main reasons was that national governments became increasingly concerned about their agricultural yields and slowly realized the economic impact of contagious diseases of livestock. The pressure on agriculture rose during serious disease outbreaks when casualties among livestock were high, so governments began to fight diseases systematically and scientifically.

In the Netherlands, the first veterinarian school was established in 1820 at Utrecht, sixty years after the first school in Europe opened its doors. In 1820, the first civil veterinarian, one Coppieters, arrived in Java. He died in 1822 and was not replaced. A decree to appoint a successor was formulated in 1838 but was withdrawn in 1839.[4] It is unclear whether more veterinarians were recruited after 1839, but until the establishment of the Civil Veterinary Service in 1853, veterinary medicine for farmers was assigned to the military horse veterinarians.[5] These veterinarians felt little responsibility for the livestock of the local population.

Because of the shortfall in manpower, the veterinary service was active only in Java from 1853 to 1869. One veterinarian served each of the three provinces, West-, Central- and East-Java. The limited number of veterinarians made it impossible to provide the whole island with veterinary care. Since the veterinarians were stationed in the capital of each province, distance and poor communications presented enormous obstacles to veterinary provision. Consequently, it was not surprising that, if called, the veterinarian often came too late and the animal had already died.

Only in 1869 were veterinarians appointed outside Java: one in Sumatra and one in Sulawesi (table 5.1).[6] The most notable increase in the number of veterinarians occurred between 1884 and 1890. Nevertheless, veterinary services did not necessarily increase since posts were often unmanned because of leave in the Netherlands or sickness or because the position remained vacant.[7] Besides, there was not much eagerness among Dutch veterinarians to work in Indonesia, and in 1914, a royal decree was issued in an attempt to recruit more staff.[8] In those days, compared to military veterinarians, those working for the civil veterinary service received low wages, worked hard, and traveled long distances. In addition, by that time, only 140 veterinarians had been graduated in the Netherlands.

A major difference compared with the Netherlands was the number of veterinarians per square kilometer, which was far lower in Indonesia.[9] Furthermore, the number of animals per veterinarian was distressingly high. With the establishment of the veterinary service, each veterinarian was responsible for approximately eighty thousand animals. After 1884, the number of veterinarians grew steadily, but the livestock population increased also. At the beginning of the 1920s, the number per veterinarian had risen to more than one hundred thousand animals, as a result of the growth in the small-stock population.[10]

Because of an inadequate number of veterinarians, hardly any fundamental research was done. This restricted any increase in basic knowl-

Table 5.1. An overview of the employees of the BVD in
Java and Indonesia in the period 1853–1938

	1853	1869	1876	1884	1890	1898	1906	1909	1913	1921	1928	1934	1938
Government vets total	3	5	7	7	26	20	22	28	35	42	43	35	32
Java	3	3					17	19	18				17
Adjunct government vets										22	42		
Indonesian vets total	—				19	17	14	9	4	3	49	58	60
Java	—			13					2				32
Military horse vets	5					8		10	13				
Paravets total	—	—	—	—	—	10	24	75		242	207	173	142
Java	—	—	—	—	—	10							79
Municipal vets[1]									4[2]				
Private vets[1]									4[3]				

1. This is the number of European veterinarians in Indonesia not in governmental service.
2. One in Surabaya, one in Jakarta, one in Semarang, and one in Medan.
3. One in Bindjai, one in Medan, and two in Surabaya.
Source: Figures compiled from the *Koloniaal Verslag.*

edge of local diseases and resulted in a slow improvement in facilities and support for the veterinary service itself. The geographical distribution of diseases was hardly known, as was revealed in the case of anthrax. In 1880, veterinarians still did not know where anthrax regularly occurred, and the first report on the disease in buffalo did not appear until 1924. Other important factors such as climate and the distances over which livestock were transported were hardly considered, so it is only logical to conclude that the veterinary service was not efficient. Although there were fewer cattle movements in Java than in the Netherlands, where serious outbreaks were often due to livestock density and animal traffic, the Dutch authorities were far more effective in repressing disease.

Lack of veterinary facilities in Java meant that local stockowners continued to use traditional medicines. The Javanese had at least one name for each disease, and livestock keepers seemed to have had an elaborate knowledge of animal diseases. Unfortunately, veterinarians rarely acquired or utilized indigenous knowledge. For example, in Lebak (residency Banten,

West-Java), cattle holders combated foot-and-mouth disease with *ayer asem*, derived from tamarind. A mixture of tamarind and salt was applied to the affected area and administered internally.[11] Another example was surra. Veterinarians were slow to accept the Javanese opinion that surra was transmitted by flies, until the link between trypanosomosis and tsetse flies had been proved in Africa.[12]

Because it was clear that the number of veterinarians was insufficient, the Dutch decided to train five Javanese. These educated Javanese were supposed to assist the European veterinarians and were called the *Indische veeartsen* (Indonesian veterinarians).[13] The experiment was not satisfactory; within nine years, only eight Javanese veterinarians had been graduated. In 1876, the school in Surabaya was closed, and the number of European veterinarians increased from five to seven.[14] Because of constant massive workloads, the government agreed to open a new school, the Nederlandsch-Indische Veeartenijkundige School (NIVS), in 1904.[15] In 1898, the first ten paravets were added to the veterinary service.[16] But in the 1920s, the institute that educated the paravets was abolished.[17]

It was not until 1888 that the first medical research institute, the Laboratory for Pathological Anatomy and Bacteriology (Laboratorium tot het doen van onderzoekingen op het gebied der pathologische anatomie en de bacteriologie), was established in Weltevreden.[18] To improve research, the Laboratory for Veterinary Research (Laboratorium voor Veeartsenijkundig Onderzoek) was erected in 1907.[19] In the same year, Melchior Treub and J. F. K. de Does established a clinic and course to educate local veterinarians.[20] According to F. C. Kraneveld, this school was an important historical development in veterinary science in Indonesia.[21] By the 1920s, eight Indonesian veterinarians were graduated each year, and some received a veterinary education in the Netherlands. Unfortunately, the economic crisis of the 1930s led to retrenchment. In 1933, no new students were accepted, and the number of teachers was reduced.[22]

An important task of the veterinary service was to study diseases that threatened economic growth. As in the Netherlands, infections such as foot-and-mouth disease, anthrax, glanders, cattle plague, scabies, surra, and rabies had to be reported to the authorities. The veterinary service also had to deal with some serious infections that were not fully understood at that time. This was because climatic conditions in Indonesia differed greatly from those in the Netherlands and also because the etiology of diseases was not fully understood since microbiology was still in its infancy. Lastly, the buffalo was an unknown animal to most state veterinarians as it was not used for agriculture in the Netherlands.

The cattle plague is discussed in this chapter in order to give an example of how the veterinary service overseas tried to improve the health of livestock. Undoubtedly, each veterinarian tried to do his best, but did government interventions really benefit Indonesian farmers? The cattle plague hit Europe hard, but why did it hit Java even harder? Was it the first outbreak of the cattle plague, or was the cattle plague already known to the Javanese?

The Cattle Plague in Europe, Asia, and Indonesia

Cattle plague has been one of the most described epizootic diseases. This is probably due to the high mortality and the speed of spread—the ingredients for a social and economic disaster.[23] In Indonesia, the earliest records appear to date back to the period 1620–1632. François Valentijn described the siege of Batavia by the sultan of Mataram, who was unable to move his artillery because a large number of buffalo succumbed to rinderpest.[24] However, the symptoms Valentijn used to describe the disease are too general for us to be absolutely sure that it actually was the cattle plague.

In the 1860s, the cattle plague hit Europe, reaching the Netherlands in 1865. Within eighteen months, almost eighty thousand cattle had died, and about forty thousand had been slaughtered. This was more than a quarter of the livestock in the infected area.[25] The disease also broke out in the 1860s in Asia, spreading through Hong Kong and Siam (1860), Calcutta and its surroundings (1864), the Andaman islands (1868), Shanghai (1872, 1875), Japan (1872), and Singapore (1874), making this epizootic an international pandemic.

In Indonesia, outbreaks of cattle plague preceded the 1860s as an epizootic broke out in Japara in 1836 and 1837 and then in the residency of Rembang between 1853 and 1879. Local impacts varied in intensity, but buffalo, cattle, and goats died in their hundreds. In 1865, a disease that at first was diagnosed as anthrax but proved to be the cattle plague raged over the west coast of Sumatra in Palembang and in Java in Tegal, Surabaya, Solo, and the Preanger Regentschappen. The cattle plague hit Madiun in 1875. The island Bawean at the north of the island Madura lost more than one-third of its livestock in 1858 because of an epidemic. In 1873, four thousand buffalo died on the island Kangean, where the government spent forty thousand guilders to buy the population new buffalo.[26]

The Cattle Plague in Java

A major epidemic designated as the cattle plague hit Java in 1879. It started in December 1878 when a dozen buffalo succumbed to an as-yet-unidentified

Figure 5.1. Administrative division of West-Java in 1884

disease at a private estate called Tjinèrè, located in Buitenzorg (see figure 5.1). On 10 January 1879, the same disease appeared on the neighboring estate, Simplicitas. This time, two veterinarians confirmed cattle plague. On 15 January, it seemed that rinderpest had spread throughout Batavia, with the center of infection being the district Parung (in the north of Buitenzorg), where approximately thirty thousand buffalo and a few thousand cows had grazed.[27]

According to the government, it was because of a shortage of ammunition, carbolic acid, and personnel that the cattle plague spread to the west in the direction of Banten. At the beginning of 1880, the whole division of Banten succumbed. In December 1879, the cattle plague was observed in Krawang, which was almost totally infected by May 1880. Then there were outbreaks in Ciancur and Sukabumi (in the Preanger Regentschappen) and Cirebon. The result was that West-Java was declared entirely infected. In June 1879, the government pronounced that, if rinderpest spread further eastward, there was a real danger of famine and the ruining of the agricultural industry.

The authorities requested assistance from the Dutch government, but this came only in 1880. The veterinary service was seriously understaffed, and besides veterinarians, many workers were needed to slaughter and bury animals as well as to enforce the quarantines.[28] Because little was known about how the disease spread, the authorities decided to erect costly fences, patrolled by soldiers, to try to contain the disease. Marines surveyed the coast of West-Java to prevent the import and export of potentially contagious cattle via sea routes. A second important measure was the slaugh-

**Table 5.2. The effects of the cattle plague
in Krawang, 1879–83**

Number of animals before the cattle plague		Percentage of animals slaughtered		Percentage of animals succumbed		Percentage left after the cattle plague	
Buffalo	Cattle	Buffalo	Cattle	Buffalo	Cattle	Buffalo	Cattle
94,297	5,674	38.6	51.7	29.8	23.1	31.6	25.4

Source: F. H. L. E. Meeng, "De veepest in Krawang," *Tijdschrift voor Nijverheid in Nederlandsch-Indië* 27 (1883): 270.

**Table 5.3. The effect of the cattle plague in northwest Cirebon
on the private estate Kandanghauer, 1881**

Number of buffalo, 1 January 1881	Died	Slaughtered	Died, slaughtered, or exported	Number of buffalo, 1 December 1881
24,065	7,343	12,038	21,081	4,052

Source: C. J. Hoogkamer, "Kritisch-historisch overzicht van de runderpest in de residentie Cheribon 1881–1882," *Veeartsenijkundige Bladen voor Nederlandsch-Indië* 10 (1897): 127.

tering of sick as well as healthy animals. Healthy animals were killed if located within a certain radius of a compound in which infected animals had been diagnosed. This measure was based on a Dutch law of 19 April 1867, which was replaced in the Netherlands on 28 July 1870 (no. 131), but not in Indonesia.[29]

The consequences were enormous, economically and sociologically. Cattle plague broke out a few times at several places, but the region under threat and the area infected differed each time. At the end of 1879, it was estimated that seventy thousand to eighty thousand buffalo, cattle, sheep, and goats had been slaughtered.[30] In Krawang 70 percent of the buffalo and 75 percent of the cattle had either died or been slaughtered between 1879 and 1883 (table 5.2). In Cirebon in Kandanghauer, the figure was over 80 percent (table 5.3). In Sukabumi, losses amounted to 5,285 animals at a cost of 443,599.05 guilders.[31] In some infected villages, over half the livestock survived, while in others, all perished. On average, around 20 percent of the original stock remained. The results were more disastrous than in the Netherlands, where rinderpest had decimated only a quarter of the national herd.

Because of the high death rate, not only the agriculture but also trade and industry suffered. During the outbreak, farmers could not use their working animals, and the relocation of livestock was forbidden. Plowing the fields became impossible, and transport links were severed. The railway company, for example, estimated that, in the Preanger Regentschappen, eight thousand animals were needed to pull four thousand carts and saw the opportunity to replace animal traction with trains.[32] The disease also had a tragic effect on individuals. In Batavia, landowners had to sell their estates, and many Indonesian farmers saw their healthy animals slaughtered. Many stockowners could not gather enough feed for their animals, which had to be kept in sheds to prevent the spread of infection and to reduce possible contact with the disease. Many animals, especially buffalo, pined away in their barns until they died. The farmers received no compensation.

The cattle plague continued for a long time. During the latter half of the nineteenth century, West-Java was never completely liberated of the disease. In May 1883, rinderpest broke out again in the residency of Batavia as a result of the importation of a herd that was being transported from Banyumas to Batavia. The herd became infected in Klari, where the cattle plague was prevalent in a mild form and therefore remained unnoticed. The transport of livestock was forbidden again in the five residencies of West-Java.[33] Banten was hit once more in 1889, and over twenty thousand animals died.[34] Further outbreaks in Banten resulted in 13,820 animals becoming infected during 1890 and 1891, of which 93 percent died. The cattle plague not only devastated Javanese herds but also spread to Sumatra in 1884.[35]

Local Knowledge, Acquired Knowledge, and Science

In 1883, it became clear that the damage was enormous, as was the pain. Unsurprisingly, the question arose as to whether the measures that had been introduced to prevent the spread of cattle plague had been appropriate: "According to the Koloniaal Verslag of 1884 the fight against the cattle plague in West-Java at the end of 1883 seemed to have cost ƒ14,039,483. Not to think about the fact that all these costs only gave us the opportunity to spread all this uncountable misery. It is gruesome!"[36]

According to a Chinese landowner Tan Yoe Hoa, who had farmed for over twenty-five years and had always possessed approximately eight hundred buffalo, the cattle plague had infected his herd before 1879.[37] Normally, 10–30 percent of his stock died, but in 1875, 57 percent perished from what Tan Yoe Hoa took to be the cattle plague. Stock on neighboring farms also died. But 1879 was the first time that the government intervened.

The list of presumed outbreaks of the cattle plague given earlier in this chapter reveals the observations and knowledge of Javanese farmers. Prior to 1879, it seems as if livestock owners had always been able to stop the disease after a certain period and a certain death rate. According to one Wirtz, an outbreak in Palembang (Sumatra) in 1865 had not made a big impression on the local population, as the memory did not last long.[38] These rapidly fading recollections suggest that the cattle plague occurred on such a regular basis that the population had become used to it.

In 1878, the government intervened for the first time and promulgated laws and decrees to combat the cattle plague. But it was these regulations that the farmer feared, lest he lose his animals, see his neighbors' cattle slaughtered, or was banned from allowing his livestock to feed or bathe. According to Tan Yoe Hoa, losses prior to government involvement amounted to around 50–60 percent, but then increased to around 80 percent.

Confirming a diagnosis was difficult because, in most cases, the veterinarian was "mustard after dinner" and "the disease had disappeared amongst the livestock."[39] Veterinarians made diagnoses with the help of cattle breeders, who did not always sketch a complete and correct picture of disease incidence.[40] Many veterinarians were unable to make the right diagnosis due to lack of knowledge or means. If the veterinarian discovered that the farmer had waited too long before requesting his services, the latter could be reprimanded. However, it was in the cattle owner's interests to keep a transmissible disease a secret as long as possible to give him time to sell, slaughter, or consume his animal(s) and prevent the total loss of his carefully gathered bovine capital.[41] If the farmer believed the disease was not contagious and would not result in death, he seldom summoned a veterinarian. Consequently, the majority of sick animals were not brought to the attention of the authorities.

State authority was also weakened by the lack of manpower, which meant that it was impossible to ensure strict compliance with the law. Many owners did not follow the regulations because they felt obedience would result in none of their animals surviving.[42] It was thus very difficult to get the support of farmers, especially as there seemed to be no individual gain and veterinarians were unable to demonstrate a shared interest between farmers and the state. There was also no effective deterrent, as offenders often escaped the fines and the veterinary service felt powerless to press charges consistently. Importantly, some veterinarians themselves had difficulties in accepting the laws:

> But even stranger are the claims, which the legislator had laid down toward disinfection of the compound or stables where sick or suspicious animals were kept. When really [if] all compounds and stables, as far as they were made from bamboo and grass and all other materials that are not susceptible to disinfection, were burned after each epizootic, then complete villages would have gone up in smoke.[43]

A major problem with the regulations was that they represented the translocation of old European laws to a very different disease environment.[44] The regulations were based on experience acquired in the Netherlands and were not necessarily suitable for the environmental, economic, and social conditions of Indonesia. The words of A. J. W. van Delden say enough: "What image, for God's sake, had the designer of this law for a Javanese village in his mind?"[45] Van Delden was critical of the inappropriateness of the slaughter-out policy as well as the areas where dead animals had to be buried, which were often too close to living quarters. Another telling fact is that, whereas Indonesian farmers had to disinfect themselves and their cattle before relocating, the external examiner, responsible for enforcing the law, did not bother to do so.[46]

When blindly introducing their laws, the Dutch authorities failed to take into account climate, types of farms, animal species, and culture. In the period 1878–83, seasonal changes caused health problems among livestock, which weakened animals and made them more susceptible to disease. Most government veterinarians were also unfamiliar with buffalo. Tan Yoe Hoa believed that the decrees had a bad impact on the health of these animals. He was convinced that cattle were three times more susceptible to cattle plague than buffalo, but that buffalo had died because of bad treatment resulting from the regulations.[47] Based on experience in Europe, some veterinarians believed that if a slaughter-out policy were introduced fast enough to an infected area, this would effectively stamp out the disease. However, there were divisions among the veterinary profession, as some practitioners, such as C. A. Penning, held the opinion that the disease could have been defeated without slaughtering a single animal; Penning blamed the external examiners for dispersing the disease, as they did not disinfect themselves and felt little compassion for the farmers.[48]

An important question that the veterinarians asked was whether the cattle plague was an indigenous or an imported disease. It was thought that, by resolving this question, answers on how to combat the cattle plague would emerge. According to Wirtz, the common view was that the cattle plague was imported to Java, either via Aceh or via Malakka.[49] Ships loaded with animals for the slaughterhouses frequently arrived in Indonesia, especially in West-Java, while East- and Central-Java purchased their slaughter cattle in Bali and Madura. Given that the incubation period for rinderpest lasts up to seven days and the journey from Singapore to Batavia was only thirty-six hours, it was certainly possible unwittingly to bring the disease into Indonesia, as happened in 1887 and 1888. If the cattle plague was discovered while at sea, the animals were slaughtered and thrown into the water.[50] Opponents of Wirtz's point of view were those who believed that the cattle plague was indigenous, and they were critical of the measures introduced to combat this disease.

It was hard to decide how to handle the cattle plague as veterinarians were unfamiliar with tropical diseases and the ideas that many infections could be spread by germs was still unproved and contested. Descriptions of diseases before the twentieth century were often rather vague, and records predating 1880 were often so concise and imprecise that it was impossible to identify the disease. Anthrax, for example, has similar symptoms to rinderpest, and veterinarians often confused the two in such a way that the cattle plague could pass unnoticed. Similarly, veterinarians were unable to differentiate between *Septicaemia haemorrhagica*, blackleg, and surra.[51] As the veterinarians were not always able to identify cattle plague, many animals were probably slaughtered when they did not actually have this notifiable disease.

CATTLE ARE sensitive to their ecological, economic, and social environment. The environment is a given setting, to which cattle holders have to adjust their strategies. The prevalence and frequency of infectious diseases were integral components of the environment. A healthy herd was of importance not only to the farmer but also to Java's economic interests. This economic dependence on livestock was one of the main reasons that the veterinary service was established in Indonesia in 1853.

However, the veterinary service struggled from the beginning due to a shortage of veterinarians, poor communications, a lack of research facilities, and the limitations of its Western training. Climate, culture, and diseases, as well as some animals, were new; as a consequence, understanding of Indonesia's disease environment developed slowly. It could be said that the veterinary service was handicapped by not having a history in Indonesia. Moreover, veterinarians made hardly any use of local knowledge. At the same time, cattle holders were reluctant to discuss infections or reveal incidence of disease because that could result in the slaughter of livestock. The official measures introduced to deal with contagious diseases did not seem to benefit livestock owners and, ironically, even some veterinarians accepted that position. Increased contact between villages and a growing economy resulted in a greater traffic in cattle, facilitating the spread of disease.

The outbreak of the cattle plague in Java in 1878 was especially significant because it was the first time that the veterinary service interfered in a contagious outbreak on a large scale. After four years, both the government and the farmers felt it had been a far greater disaster than had been anticipated at the beginning. For the government, the price paid in terms of controlling the disease, as well as indirect losses in trade in agricultural commodities, seemed far too high. For farmers, the psychological and physical suffering they endured through their stock deaths damaged their livelihoods and undermined their farming systems.

According to the sources, the cattle plague was already known in Java before 1879. This meant that the government paid fourteen million guilders for something it did not have to control because local farmers already knew how to handle this disease. It appears that, as a result of interference by the state veterinary service, the cattle plague spread rapidly over an enormous area. In Java's history, such an outbreak with such a high death rate had never been recorded before.

Nevertheless, it is impossible to measure definitely the exact effect the regulations had on the livestock economy. However, it is clear that it was impossible to inspect an area as large as West-Java with such a small number of veterinarians and their assistants. The inability to promulgate and then ensure compliance with the decrees, the use of laborers who did not feel connected to the problem, and the veterinarians' own lack of confidence led to an inefficient battle to contain the epizootic. It is also unclear the extent to which defiance of the law or inappropriate measures actually aggravated the situation. On the one hand, it seems logical to accept the fact that the cattle plague was already known before 1878 because local farmers had a variety of names for the disease and they seemed to have their own ways of dealing it. On the other hand, it was also possible that the seriousness of this out-break was due to a more violent strain of the virus, so that the extent of this catastrophe had nothing to do with government interference. Lack of scientific data on this and past epizootics makes this fact impossible to ascertain. Overall, however, the historical sources suggest that state interference in the 1878 epizootic exacerbated the deleterious impact of this outbreak. The actions of the civil veterinary service cost the farmers of the nineteenth century a lot, both in terms of money and physical and emotional pain.

Notes

1. Lise Wilkinson, "Rinderpest and Mainstream Infectious Disease Concepts in the Eighteenth Century," *Medical History* 28, no. 2 (1984): 141.

2. M. W. Barwegen, "Gouden hoorns: De geschiedenis van de veehouderij op Java, 1850–2000" (PhD diss., University of Wageningen, 2005).

3. W. Groeneveld, "Het paard in Nederlandsch-Indië: Hoe het is ontstaan, hoe het is en hoe het kan worden," *Veeartsenijkundige Bladen voor Nederlandsch-Indië* 28 (1916): 228. All translations are my own.

4. Governmental decree (GD) of 13 January 1838, no. 19, and GD of 10 August 1839, no. 7.

5. J. Boeke, "Veeteelt," in *Gedenkboek voor Nederlandsch-Indië: Ter gelegenheid van het regeeringsjubileum van H. M. de Koningin, 1898–1923,* ed. L. F. van Gent, W. A. Penard, and D. A. Rinkes (Batavia/Leiden: Kolff, 1923), 332; F. C. Kraneveld, "Veterinaire varia van Indonesië No IV: Een flitsbeeld over de geschiedenis der diergeneeskunde van Nederlands Oost Indië gedurende de periode 1820–1940," *Hemera Zoa* 65 (1958): 96–113.

6. GD of 25 October 1869, no. 8.

7. *Jaarboek Landbouw,* 1912, 259; *Verslag BVD,* 1913, 1; *Verslag BVD,* 1914, 4.

8. *Jaarboek Landbouw,* 1909, 263; idem, 1915, 217; S. R. Numans, *Het probleem der klinische opleiding van de dierenarts in Indonesië in verband met zijn taak in de maatschappij* (Groningen [etc.]: Wolters, 1951) [Speech], 8; Royal decree (RD, "Koninklijk Besluit") 25 June 1914, no. 60 (*Indisch Staatsblad,* 1914, no. 697).

9. Java (including Madura) has a surface area of 130,000 km², whereas the Netherlands comprises only 35,000 km². H. E. B. Schmalhausen, "Waaraan heeft Java behoefte?" *De Indische Gids* 24, no. 2 (1902): 1047; F. C. Kraneveld, *Bestrijding van dierziekten in de tropen* (Utrecht: Drukkerij v.h. Kemink en Zoon N.V., 1949).

10. In 1853, the number of small livestock was not yet counted.

11. D. J. Fischer, "Is pyotanine nog niet geëvenaard als geneesmiddel tegen mond- en klauwzeer?" *Veeartsenijkundige Bladen voor Nederlandsch-Indië* 6 (1892): 189–92.

12. J. K. F. de Does, "Bijdrage tot de kennis der trypanosomenziekten, in het bizonder die, welke op Java voorkomen," *Veeartsenijkundige Bladen voor Nederlandsch-Indië* 13 (1902): 313–47.

13. *Indisch Staatsblad,* 1860, no. 1.

14. *Indisch Staatsblad,* 1876, no. 348.

15. The NIVS is now the Faculty of Veterinary Medicine and Animal Science in Bogor, which is part of the IPB, the Bogor Agricultural University.

16. *Indisch Staatsblad,* 1898, nos. 80 and 314.

17. J. Ballot, "Historisch Overzicht van de Maatregelen tot Verbetering van den Paarden- en Veestapel in Nederlandsch-Indië," *Veeartsenijkundige Bladen voor Nederlandsch-Indië* 10 (1897): 17–87.

18. This later became the Eijkman Institute.

19. *Indisch Staatsblad,* 1907, no. 89.

20. GD of 24 January 1907, no. 63 (*Indisch Staatsblad,* 1907, no. 66). Melchior Treub was at that time the director of the Ministry of Agriculture, Trade, and Industry. De Does was a veterinarian who wrote many scientific publications.

He was attached to the Laboratory for Pathological Anatomy and Bacteriology from 1896 to 1907 and was one of the main founders of the Laboratory for Veterinary Research.

21. Kraneveld, "Veterinaire varia van Indonesië No IV," 98–100. Kraneveld was the director of the Instituut voor Tropische en Protozaire Ziekte at the Faculty of Veterinary Science when he wrote this article. Before World War II, he had been a veterinarian in the Netherlands-Indies.

22. *Indisch Staatsblad,* 1919, no. 359; and 1920, no. 58.

23. J. Blancou, *History of the Surveillance and Control of Transmissible Animal Diseases* (Paris: Office International des Épizooties, 2003).

24. D. Driessen, *Ueber die Tierseuchen besonders über die Rinderpest in Niederländische Ostindien* (Venlo: Uyttenbroeck, 1885), 8.

25. J. Bieleman, *Geschiedenis van de landbouw in Nederland, 1500–1950* (Meppel, Netherlands: Boom, 1992), 291; Van der Kemp, "Een terugblik op de over West-Java gewoed hebbende runderepizoöthie," *Tijdschrift voor Nijverheid en Landbouw in Nederlandsch-Indië* 30 (1885): 139.

26. Driessen, *Ueber die Tierseuchen* 8–12.

27. Van der Kemp, "Een terugblik op de over West-Java," 188–89.

28. Ibid., 140–44.

29. D. Driessen, "Bijdrage tot de runderpest-geographie," *Geneeskundig tijdschrift voor Nederlandsch-Indië* 21 (1881): 309; Van der Kemp, "Een terugblik op de over West-Java," 265–89.

30. Arsip Nasional Republik of Indonesia, Jakarta, appendix to the decree of 5 June 1880, no. 1.

31. Arsip Nasional Republik of Indonesia, Jakarta, appendix to the decree of 14 March 1882.

32. X., "Veepest en staatsspoorwegen op Java," *De Indische Gids* 2 (1881): 455–57.

33. GD of 25 May 1883, no. 25.

34. C. A. Penning "Runderpest in Ned.-Indië," *Veeartsenijkundige Bladen voor Nederlandsch-Indië* 6 (1892): 41; Driessen, *Ueber die Tierseuchen,* 17.

35. In April 1887, the cattle plague appeared in Palembang, in the southeast of Borneo. From June until October 1887, the cattle plague raged in Pontianak, Montrado, and Batavia. In December 1887, the Siamese slaughtered animals, and in April 1888, a herd of Siamese cattle in Tanjung Priok was contaminated. In 1888, the cattle plague claimed another 7,994 victims in Batavia (Penning, "Runderpest in Ned.-Indië," 41).

36. Van der Kemp, "Een terugblik op de over West-Java," 306.

37. J. M. van Vleuten, "Advies van een Chineeschen landeigenaar in zake veepest," *Tijdschrift voor Nijverheid in Nederlandsch-Indië* 28 (1884): 149–63.

38. Wirtz, "Rapport," *Tijdschrift voor Nijverheid in Nederlandsch-Indië* 27 (1883): 497–522. At the time Wirtz wrote this article, he was the director of Veterinary School (not yet a university) in Utrecht.

39. Driessen, "Bijdrage tot de runderpest-geographie," 318–19; Fischer, "Is pyotanine nog niet geëvenaard," 153; G. Ad. van Lier, "Eenige mededeelingen over surra," *Veeartsenijkundige Bladen voor Nederlandsch-Indië* 17 (1905): 239; J. Ch. F. Sohns, "B. Boutvuur in Nederlandsch-Indië," *Veeartsenijkundige Bladen voor Nederlandsch-Indië* 26 (1914): 254.

40. Sohns, "B. Boutvuur in Nederlandsch-Indië," 257.

41. Van Eecke, "Pectorale vorm van *Septichaemia haemorrhagica*," *Veeartsenijkundige Bladen voor Nederlandsch-Indië* 9 (1895): 300–6.

42. Van Vleuten, "Advies van een Chineeschen landeigenaar."

43. C. S. Jeronimus, "Mond- en klauwzeerbestrijding in Indië," *Veeartsenijkundige Bladen voor Nederlandsch-Indië* 29 (1917): 20.

44. J. Breedveld, "Over mond- en klauwzeer," *Veeartsenijkundige Bladen voor Nederlandsch-Indië* 14 (1902): 245–48.

45. Quoted in Van der Kemp, "Een terugblik op de over West-Java," 354. A. J. W. van Delden was a private cattle owner who kept an archive on the cattle plague.

46. Driessen, "Bijdrage tot de runderpest-geographie," 309; Van der Kemp, "Een terugblik op de over West-Java," 265–289.

47. Van Vleuten, "Advies van een Chineeschen landeigenaar," 159.

48. Penning, "Runderpest in Ned.-Indië," 29; F. H. L. E. Meeng, "De veepest in Krawang," *Tijdschrift voor Nijverheid en Landbouw in Nederlandsch-Indië* 27 (1883): 265–328; Van der Kemp, "Een terugblik op de over West-Java."

49. Wirtz, "Rapport," 504

50. Penning, "Runderpest in Ned.-Indië," 35–37.

51. W. J. Esser, "In zake v. Eecke's onderzoekingen omtrent het voorkomen van *Septichaemie haemorrhagica* en runderpest, speciaal omtrent oedemateuze runderpest, onder den veestapel in Nederlandsch-Indië," *Veeartsenijkundige Bladen voor Nederlandsch-Indië* 6 (1892): 71–72; G. Ad. van Lier, "Eenige mededeelingen over surra" *Veeartsenijkundige Bladen voor Nederlandsch Indië* 17 (1905): 213–40; C. A. Penning, *Trypanosomosen in Ned.-Indië* (Semarang-Soerabaia: G.C.T. van Dorp & Co., Vereeniging tot Bevordering van de Veeartsenijkunde in Nederlandsch-Indië, 1904).

CHAPTER 6

Fighting Rinderpest in the Philippines, 1886–1941

Daniel F. Doeppers

THE GREAT epizootic waves of rinderpest that devastated Philippine bovine populations in the late nineteenth and early twentieth centuries are the focus of this chapter.[1] This disease struck not only cattle (*Bos taurus*) in great numbers but also water buffalo, or carabao (*Bubalus bubalis*)—the essential work animal in Philippine wet-rice agriculture. Total provincial bovine loss rates of around 85 percent were recorded in the first two waves. Such catastrophes often left local rice fields unworked for years afterward. First, the changing dynamics of the Philippine cattle- and sheep-importing business is considered, since this offers the best explanation of how and when rinderpest was transferred from the Asian mainland to the archipelago. Then, the unfolding geographies of the three great rinderpest epizootics are sketched. Finally, the actions taken and not taken by the authorities in their long frustrating but ultimately successful effort to limit livestock mortality are reviewed.[2]

   With the advent of mass-market hamburgers, beef eating has become commonplace in Manila. As recently as the 1960s, however, beef was not a significant part of the ordinary, less-affluent Filipino's diet, and a great many Manilans were not used even to tasting beef. For a long time in the nineteenth and early twentieth centuries, the meat from cattle slaughtered in the city was consumed mainly by Spaniards and other foreigners and wealthy cosmopolitan Filipinos. Some Spaniards in the 1890s even advocated a heavily meat—meaning beef—diet as a protection for European constitutions in the tropics. For affluent Filipinos, it would have been a mark of class. Carabao meat, or "carabeef" as it is sometimes called today, was eaten on occasion in the city. But there was a certain prejudice against it, and there were often official regulations aimed at preserving the carabao population for agricultural and draft purposes. From 1925 to 1933, only about 2 percent of the bovines slaughtered annually in Manila for human consumption were carabao.

Domestic cattle are an introduced species in most parts of insular Southeast Asia and not of deep antiquity. The major exception is found in part of Indonesia, where Bali cattle (and the original Java cattle) were domesticated from the native *banteng* (*Bos javanicus*). In the northern and central portions of the Philippine archipelago, cattle were introduced under Spanish aegis in the sixteenth century from China, Mexico, and Spain. The introduction of cattle to the insular Southeast Asian lowlands has been a long and incomplete process.[3] In any case, for many years, only a few Filipinos consumed beef with any regularity—and when they used milk, it usually came from a carabao.

Imported Animals and Epizootic Disease

Caused by a virus, rinderpest attacks the mucous membranes of the body, especially the digestive tract.[4] High fever, ulcers of the membranes, dysentery, and death in a week or less are typical. In general, cloven-hoofed ruminants are susceptible at one level or another. Rinderpest is readily transmitted by close association with an infected animal through contact with nasal and other discharges, dung, and/or urine. The virus may be transferred directly or indirectly through food. In Shanghai, dairy cattle got the disease when they were fed fresh cotton-seed cake believed to have been contaminated by infected animals working in the local cotton mills. Rinderpest is less likely or unlikely to spread through the air or by insect transmission, and even direct discharges are believed to lose their virulence following two days in sunlight.[5]

Although equivalent patterns for the whole of Southeast Asia remain to be worked out, there was a significant trade in animals in many areas in the late nineteenth century, and these live-animal flows often resulted in disease transmission. Siam/Thailand, in particular, was annually exporting thousands of bullocks to Singapore and Sumatra in the 1880s and 1890s. This trade crashed in 1897 when "rinderpest . . . ravaged the whole of central Siam, attacking both buffalo and oxen with such severity that the [rice] harvest prospects [were] seriously threatened." At the same time, work and milk animals were also exported from India to Malaya and Singapore. Singapore maintained an open-import policy on livestock until an outbreak of rinderpest in Calcutta in 1935 threatened dire consequences. Until then, milk animals were imported directly from the Punjab, passing by rail to Calcutta and by sea to Singapore. In nearby Indonesia, according to Martine Barwegen, there was a major rinderpest event in West-Java starting in 1879 and others later.[6] Looking further afield, animal imports led to massive epizootics in East and Southern Africa in 1897, with greater than 90 percent mortality in Swaziland and Natal. Rinderpest was already well known in India and Egypt. Indeed, in the Malay Peninsula, adult water buffalo imported from India proved more resistant than local stock precisely because they had already been exposed to a disease-rich environment of which rinderpest was a part.[7] Even the present slender research record for Southeast Asia shows that the diffusion of rinderpest, anthrax, and other bovine diseases was a common consequence of the quickening pace of commerce in animals during the late nineteenth century.

Toward the end of the nineteenth century, rinderpest devastated the Philippines. Presumably this disease entered the country because the colonial government approved the idea of providing more fresh meat other than fish or pork for foreign nationals, especially for Spaniards. Imported animals and epizootic disease became critical to the meat supply of the city after 1886. Just as the archipelago lost its surplus position in rice in the 1870s, so it lost self-sufficiency in bovines a decade later. More than the loss of local beef supplies, the loss of most work animals for rice production was broadly significant to human welfare and the national economy for a decade or so following each epizootic wave.

Two new trends in meat provisionment emerged around 1870. The number of Spaniards in Manila and the archipelago began to expand sharply, increasing the demand for beef. During the 1870s and early 1880s, the same can be said of the well-educated mestizo elite. Further, there were growing pressures for settlement and crop production that tended to diminish the available area for open-range cattle production in the more accessible parts of Luzon. Whereas in the 1850s, a beef cow was often sold for less than a large hog, after about 1860, cattle owners enjoyed some increased valuation on their stock.

Rinderpest entered the country in the 1880s presumably because the colonial government allowed private dealers to provide more fresh beef for foreign nationals or because it entered with shipments of sheep for the tiny mutton trade. Some animals were also imported to work on the growing sugar estates. Alternatively, the disease may have entered via a few water buffalo brought for breeding. In any case, a few imported animals led to a world of pain.

It is just possible that rinderpest arrived in Luzon via the regular mutton trade. Though not a prominent part of the indigenous diet, mutton was certainly consumed by Manila's European population, and it was consumed also by the small number of urban residents from the Middle East. Some of the more affluent indigenous inhabitants also developed a taste for it.[8] Sheep did not flourish in the archipelago, but for decades, a few sheep were slaughtered in the public abattoir each week. These animals came from abroad. The numbers of "sheep" (*carneros* or *ganado lanar*) range at first from scores annually to hundreds by 1866. Significantly, 2,398 *animales lanar* were imported to Manila during the three years 1884–1886—more than two thousand of these were slaughtered for the city's public markets in the same years. The first rinderpest outbreak to attain epizootic status began in 1886. As the losses from disease escalated, a fresh-meat deficit was created, and the rate of animal imports accelerated. Almost thirteen hundred sheep were imported in 1890. Thereafter, the import category is relabeled "sheep and goats," and the numbers jump to sixty-four hundred and ninety-eight hundred in 1891 and 1894, respectively. In the 1880s, these animals came from "British possessions," presumably Hong Kong, and in the 1890s, from "China"—now statistically including Hong Kong. In 1891, almost thirty-five hundred of the imported sheep came from China.[9] The relevant point is that small ruminants, sheep or goats, can be infected with rinderpest, although in general they are susceptible at lower rates than cattle. While the disease is active in individual animals, though uncommon, it can be transmitted from sheep to cattle under direct contact as in a corral or pen via nasal discharges and excretions.[10]

Not before 1884 does the import category "bovine" (*animales vacunos*) yield a significant number, with sixty arriving from Hong Kong and eight from Australia. Given the disease environment in and about the point of origin, these few bovine imports from the China Coast are highly likely to have included multiple animals infected with serious disease. Although most were surely intended for slaughter, they could well have been held in private corrals in the city where animals ordinarily mixed without aggressive quarantine. A few would likely have been sold to buyers from nearby provinces. It is clear in retrospect that the authorities were not on the lookout for this disease.

Whether sheep or cattle or even a few carabao, in the end, this modest flow of live animals resulted in the introduction of disease. Given the right situation of direct contact, it takes only one infected animal. This seems the most reasonable reading of the available evidence. It is possible, of course, that rinderpest entered the archipelago at some earlier point and was not recognized or officially noted because it remained localized.[11] In the Philippine archipelago, three waves of lethal rinderpest epizootics ensued, peaking in the late 1880s, the turn of the century, and the later 1910s to the early 1920s. The resulting mortality of major animals created a dire need to import more for work and slaughter. This created bonanza opportunities for import-stock dealers, and they in turn proved politically and financially adept at becoming a factor in the formation of state policy on further imports. The rinderpest epizootics were devastating to the livestock of the archipelago, to the farmers who used or raised carabao and cattle, and to the larger economy due to the cost of rice imported to replace lost domestic production.

Initial outbreaks with 85 to 90 percent mortality among cattle are entirely typical of the disease called "cattle plague" in England and "rinderpest" in many other places. This was a very old disease in Europe with outbreaks in the seventeenth and eighteenth centuries associated with the oxen trade between Denmark, the north German lands, and the Netherlands. The notion that it was a "state" function to take action to stop these epizootics dates from at least this time, as do the ideas of quarantine, barriers to trade and movement, animal-health certificates, the mass slaughter of infected and exposed animals, and even policies aimed at recovery and restocking in affected areas.[12] Late outbreaks in Europe tended to follow in the wake of the disrupted conditions associated with wars: the Napoleonic Wars, the Franco-Prussian War of 1870, the Balkan Wars in 1913, and in Poland at the end of World War I. The disease was eradicated in the Netherlands only in 1866 and England in the 1870s—using strong state methods of quarantine and regional mass slaughter. Neither Spain nor its Pacific dependency was a strong state.

The First Wave in the Philippines

In the late nineteenth century, rinderpest was endemic in the hinterlands of all of the great ports of the China Coast, including Hong Kong. It is likely the same was true of southern Indochina. The first wave in the Philippines almost certainly may be traced to the animals imported from Hong Kong. The disease reached crisis proportions in the Philippines during late 1886 and 1887. Dutch Consul Hens wrote that conditions in several nearby provinces were already severely disturbed:

At the highest point of these problems . . . an epizootic broke
out for six months that killed two-thirds of the farmers' beasts,
especially buffaloes and cattle and the government couldn't do
anything to stop it. The cadavers infested the air and the rivers,
we bless providence that the epidemic didn't attack our species.

Hens reports that the epizootic started to the east and southeast of
the city and then seemed "to follow the wind of the southeast monsoon,
to stop, we hope, in the northwestern provinces of Luzon on the China
Sea."[13] This hope proved illusory. Hens's account of the early track of the
epizootic is almost exactly confirmed by the dramatic decline in shipments
of slaughter cattle to Manila first from Laguna (southeast of the city) and
then Pangasinan, and subsequently Ilocos (both to the north), in that order. Cattle from Laguna slaughtered in Manila declined from 1,711 in 1885 to
328 in 1886 and 28 in 1887. The progress of the disease eliminated this province as an important supplier of meat to the city. The military veterinarian
Gines Geis Gotzens states that the epizootic tended to follow the lines of
commerce and communications. These lines ran up the west side of the
Central Plain, and shipments from Nueva Ecija on the east side of the plain
did not decline during 1887. By the end of that year, what was subsequently
understood to have been rinderpest had spread from the Marikina Valley
just east of the city and Bulacan just northwest, blanketed parts of Central
Luzon, and entered the Ilocos coast and Nueva Vizcaya, both to the far
north. Cattle shipments from Pangasinan for slaughter in Manila declined
from more than forty-five hundred in 1885 to fewer than two thousand in
1887. As even this diminished flow of animals moved overland, it further
spread the disease. From Pangasinan, rinderpest was transmitted west into
the Bolinao Peninsula, devastating the livestock of Alaminos in 1888.

The epizootic also spread south of Manila to Batangas, Tayabas, and
Cavite provinces, though there was no immediate decline in shipments.
Indeed, cattle coming to the city from Batangas and especially Tayabas increased during 1887, picking up the slack in supply caused by the decline
from other sources. In Batangas, many cattle were raised singly or in small
groups and were not allowed to run free. This practice may have slowed the
progress of the disease in that province. Rinderpest was recorded in central
Batangas at Rosario in the period 1887–88. More than seven hundred cows
died in this wave in Santa Cruz in neighboring Cavite—now called Tanza.
During the following year, Ilocos was further devastated. By 1889, the worst
of the first wave was playing itself out in the Cagayan Valley in northeastern
Luzon. Rinderpest was also spread by sea to Iloilo and Capiz provinces on

Panay. In the face of this, the domestic supply of slaughter animals to the city was not maintained.[14] By mid-1888, the supply was insufficient, and in the following year, the total annual slaughter of beef cattle in the Manila abattoir had fallen from around 21,000 in 1886 and 1887 to 15,700. The local slaughter in Manila continued well below normal in the early 1890s, falling below 15,500 in 1894. There were places now where "hardly any carabao or cattle were left alive."[15]

The response of the colonial government was slow and uncertain—further evidence of a lack of familiarity with the disease. A major circular of regulations aimed at combating its spread was issued by the Inspeccion de Benefiencia y Sanidad in October 1888 and was renewed at the end of 1890. Late in the cycle, these regulations attempted to impede the transport of diseased animals and required "scrupulous vigilance" at the slaughter-houses to keep such animals out of the food supply. The regulations provided for a fifteen-day quarantine of suspicious animals in areas where the disease had broken out and recommended keeping these animals from contact with goats, dogs, pigs, and other animals that ran loose and were thought capable of spreading the disease. It recommended disinfection procedures as well as the cremation of animals that died from the disease. These regulations may have helped, given that there was a stable administrative system in place to carry some of them out, but they came too late to seriously impede diffusion in Central Luzon. They may have helped to protect the livestock of places less intensely integrated with the city. In the end, the impact of this epizootic was not universal, and some areas escaped. In a subsequent review of the Philippine evidence, a veterinary wrote of the first wave, "the disease must have run a sporadic and mild course after the first severe onset."[16]

What the regulations did not mention was any proactive slaughter of diseased or likely-to-have-been-exposed animals in the vicinity of the outbreaks, a major weapon in the arsenal of animal-disease control. It was used extensively in the Netherlands and United Kingdom in their efforts to eradicate rinderpest twenty years earlier. It was also used by Dutch officials in combating rinderpest in Java from 1879 to 1883. This may be what Consul Hens had in mind in writing his report on 1887, cited above: "an epizootic broke out for six months that killed two-thirds of the farmers' beasts, especially buffaloes and cattle and the government couldn't do anything to stop it." That is, he thought the Spanish government response was too slow and soft, not at all the sort of strong-state response with which he was familiar. Of course, this nonpolicy avoided the possibility of making things even worse and/or provoking a backlash. In the Philippines, slaughter of "sick and exposed animals, with a certain amount of indemnity" was tried briefly in 1911 in the aftermath of the second wave, but was discontinued because of strong farmer opposition and their attempts to hide sick animals. Subsequent American and Filipino authorities largely eschewed this weapon.[17]

In portions of central and northern Luzon, this epizootic wave became a disaster of the first moment. Foreman reports a stockowner in Bulacan who lost 85 percent of his animals. Ken De Bevoise cites data indicating the loss of at least 84 percent of carabao and cattle in Pangasinan Province, and Paul Rodell writes that this epizootic wave "almost completely destroyed the Zambales cattle industry," with the reported numbers declining from more than 23,000 in 1886 to fewer than 2,900 in 1892. In the northwest peninsula, the important stock raising communities of Bolinao and Anda fared even worse than Zambales as a whole, with cattle declining from a combined 9,660 to approximately 385. To the residents of Alaminos in western Pangasinan with losses of 8,000 cattle and 5,000 carabao, it was the end of a stock-raising era. Numbers of farm carabao also declined across Zambales, but less drastically—probably because they were more isolated from other bovines. Since range cattle raising was a major form of commercial adaptation in the hilly lands of Zambales and the Bolinao peninsula, one can just guess at the economic hardship unleashed. Other areas were less impacted. In Bikol (southern Luzon) and much of the Visayas outside Panay, the first rinderpest epizootic seems to have been far less significant than the second.[18]

The Second Wave

The incidence of new cases did not decline to zero, but still the numbers of carabao and cattle in the archipelago were gradually rebuilt. Work animals do not seem to have been in notably short supply in 1895 and 1896, just prior to the outbreak of the Philippine Revolution. Still, because carabao mothers are careful nurturers of calves, their numbers would ordinarily have taken longer to rebound than those of cattle. Also, in the late 1890s, extremely cheap cattle from Queensland were imported, presumably in some numbers, and became a likely source of disease. In any case, a little more than a decade later, there were sufficient animals born since the first wave to sustain another. Although animals born to rinderpest survivors would initially have had some immunity to the disease, in general, this immunity lasts less than a year.[19] The second wave was made much worse by the general disruption of the country caused by the collapse of administrative authority in face of the revolution and, more particularly, by its coincidence with the invasion of American forces who moved their troops and draft animals around the archipelago without regard for the disease. It was some time before new local and provincial authorities were firmly in place. One cannot be sure how much of the severity of the second wave can be ascribed to the special conditions created by warfare, but clearly these conditions exacerbated the situation. Ken De Bevoise points to the likelihood of increased rinderpest transmission as a result of war refugees' taking surviving carabao into concentrated and unsettled conditions and because the American army actively used requisitioned carabao for military transport. All this fits the pattern of conditions leading to the rinderpest outbreaks in Europe in the nineteenth and twentieth centuries. As a result, this wave of the disease had a more uniformly devastating impact around Manila and in Central Luzon than the first. Subsequent veterinary authorities thought that a "continual intermingling of animals" in the inner lowland plain of Central Luzon led to a severe rinderpest impact. Considerable rice land was still out of production there in 1908 and even later for lack of plow animals.[20]

In De Bevoise's reconstruction, the second cycle began in 1898 or 1899 in the Southern Tagalog areas most affected by the revolution and spread outward devastating Central Luzon in 1899. Ultimately, in 1900 and 1901, very large numbers of bovines died in the extremities of the island to the far north in the Cagayan Valley and to the far south in Bikol—again with reports of great riverside tangles of rotting carabao corpses. Norman Owen flatly states that "the rinderpest epidemic of 1900 . . . virtually destroyed the local cattle industry" in Bikol. The disease was now claiming many victims outside Luzon in Marinduque, Leyte, and even northern Mindanao. A retrospective census question on cattle and carabao mortality during 1902 resulted in a reported figure of 629,000 animals having died during that year in the major portions of the archipelago covered by the census, as against only 80,000 slaughtered for meat. In that year, late in the second cycle, the greatest mortality concentrations were in the Central and Western Visayas. Negros, Bohol, Cebu, Iloilo, and Leyte each reported more than 50,000 dead. Only a far-flung scattering of isolated places seemed to have escaped into 1903. The great dip in the cattle-to-human population ratios for most provinces stands as mute testimony to the devastation caused by this epizootic. Beyond the purview of this treatment is the relationship between the mass death of bovines and the human-health effect, as malaria-transmitting mosquitoes, deprived of their favored large animal targets, turned increasingly to prey on people.[21]

With the economy of the rice-producing areas in collapse and famine abroad by 1902, many families resorted to roots and tubers for subsistence.

The government attempted to speed the replacement of work animals by purchasing and importing them from abroad. During 1903, an estimated thirty-five thousand carabao were purchased in China and brought in, with more following in 1904 and 1905, but many of these died of disease. Acquiring animals and getting them distributed in good health without more effective disease control was an impossibility. In the receiving provinces, one example can stand for many. In Zambales in late 1903, the governor reported that, despite the lack of work animals, few people were willing to try out carabao imported from China. Finally, a landowner in Iba bought some imported cattle, "but they died of rinderpest a few days after their arrival and spread contagion to his [remaining] carabaos and then rapidly to other animals in the locality." Diffusion of this outbreak was stopped when adjacent municipalities prevented the movement of animals into their jurisdictions.[22] Still, the market for live animals for meat and work continued to reward dealer-importers, despite the health and economic impact of continued disease transmission. Rice agriculture in muddy pond fields required bovine work animals.

Imports were wide open. Almost twenty thousand head of live bovines arrived from Hong Kong during fiscal 1907–8, with lesser numbers coming from Xiamen (Amoy), Hainan, and Taiwan. The animals were still being brought into the city and country and just as surely bringing disease with them. The Australian trade commissioner claimed that stock from China formed the largest block of imports because Chinese animals were quiet and easily handled—as opposed to the demeanor of Australian range cattle. Just as certainly, low cost was a major factor.[23]

The other major bovine supply zone in the first decade of the new century was Indochina, especially Cambodia and the river port of Phnom Penh and secondarily Saigon and Vinh in Vietnam. Although Indochina was the source of ninety "live animals" in both 1877 and 1880 and fifteen bovines in 1891, there is little direct evidence that it was a major supplier to the Manila market until just before the turn of the century. Nevertheless, the Philippines became the major outlet for cattle from Cambodia and Vietnam. From the Cambodian side, total exports of cattle peaked in 1898 and 1899, coinciding perfectly with the start of the second rinderpest wave in the Philippines; plummeted by half during a period coinciding with the Philippine-American War; and then recovered sharply during 1911–13, surpassing the peak of 1899 in the last year. The destination of the exported animals is not precisely identified, but the Philippines was the major market for Cambodia. In the Philippines, some 16,600 bovines arrived from Indochina in fiscal 1907–8.

Although the animals were held in corrals in Phnom Penh and Saigon pending shipment or in small lots along the river awaiting transfer to a steamer, they were not routinely subject to inspection by veterinarians as late as 1908. Nor were the French authorities ignorant of the disease. Even when thus quarantined, resistant animals from areas where rinderpest was endemic might well display only subtle symptoms. In one shipment from Indochina in mid-1907, 100 out of 375 animals were found to be suffering from rinderpest by the time the ship arrived in Manila.[24]

The major buyer-importers were Filipinos. Faustino Lichauco came to be labeled the "cattle king of the Philippines." He derived a considerable income from this business and in his time was known for a large and stylish household, lavish social and political entertaining, and the numerous extended residences of his wife and children in various countries of Europe and the United States. His first cousin and business rival, Ramon Soriano, also entered this business, importing hundreds of cattle from the Chinese port of Hoihow in 1908. Another importer was the British trading and management company known as Smith Bell bringing in cattle from Phnom Penh in 1910.[25]

Having made little progress in the biological control of rinderpest during the first decade of the twentieth century, the livestock division of the Philippine Bureau of Agriculture renewed its efforts at geographical control. After all, the cattle plague had been effectively combated in western Europe by aggressive use of quarantine, segregation, and slaughter. Since bovine diseases were readily spread by cattle trading, to say nothing of unfenced pasturing, the Bureau of Agriculture now attempted an internal quarantine. Lacking personnel even remotely adequate to the task at the national scale, bureau executives decided to concentrate on imposing a strict animal quarantine at the northern end of the Central Plain in the province of Pangasinan beginning in 1911. Pangasinan was chosen because it was a critical rice-producing province and because rinderpest was again spreading there. Further, the province lay astride strategic choke points in the routes south into the Central Plain from both the Ilocos coast and the Cagayan Valley. These routes, especially that down the coast, were being used by migrating Ilocano rice farmers seeking to settle along the railway lines and by dealers seeking the high prices available for work animals in the reviving sugar industry of Pampanga. Government veterinaries believed

the hill country of Pangasinan [the Bolinao Peninsula], which on account of its rough mountainous character is essentially pastoral instead of agricultural, and the Ilocos Provinces have not been affected by rinderpest as severely as the central valley. The reason for this is that on account of the rough nature of the country, the very poor roads, and the proximity to the sea, the majority of products are transported by water. Therefore, no such continual intermingling of animals occurs as in the central [plain].[26]

Thus, east-central Pangasinan—the prime intermingling point—was a well-chosen place to start. The army agreed to cooperate in the quarantine effort by providing twelve hundred scouts, some cavalrymen, and five veterinarians. The Customs Bureau also cooperated by banning the transport of carabao and cattle in small boats and requiring health certificates for movement on larger boats. Quarantine stations were established along the major land routes: at Aringay in La Union, Camp One on the Baguio Road in Benguet, San Nicolas in Pangasinan, and Carranglan in Nueva Ecija, the last two on trails leading southwestward from Nueva Vizcaya. A barbed wire cordon sanitaire was constructed along the Pangasinan–La Union border. As a result of this concentrated effort, rinderpest was gradually eradicated from the eastern two-thirds of Pangasinan and thereby impeded from further diffusion southward into the heart of the Central Plain. The gains seemed promising, but in the general absence of fencing, even one infected animal entering after the military quarantine was withdrawn could undo it all. At the end of fiscal 1911, at least eighty-one municipalities across the archipelago reported active cases of rinderpest.[27] The ban on movement, use of health certificates, and specific quarantines were all techniques pioneered across northern Europe centuries earlier.

The Philippine authorities were in a bind. One the one hand, the shortages due to the ravages of disease created an immediate need to import cattle for slaughter and, of course, for agricultural work. On the other hand, the import of animals from foreign areas unprepared to provide disease-free stock meant the frequent reintroduction of lethal disease vectors. Relatively affluent and influential urbanites, to say nothing of the U.S. Army, wanted beef, and the well-connected importers wanted the continued opportunity for profit in this commerce. Both Lichauco and Soriano were landing large numbers of infected cattle, but both were politically well connected. As a result, the Bureau of Agriculture made concessions to them. One concession was waiting until the third shipment of cattle arrived in a highly diseased state before declaring the partner port quarantined. The technocratic authorities would have been happy to ban imports of live animals as a way to stop the devastation of continual reinfection. Even in the absence of effective medical therapies or prophylaxis, one could not simply dismiss the evidence of continued disease transmission. Given these countervailing pressures, what was possible was to continue the work of public education and to institute a quarantine system based at first on the Sanitary Code of the City of Manila. This system would intercept diseased animals coming from abroad and attempt to reduce their opportunity to ignite another epizootic. This was vigorously opposed by some local cattle dealers.[28]

As part of this effort, a quarantine station was constructed in the Pandacan district of the city accessible to international vessels via lighters on the Pasig River. Meat animals passing final inspection there were walked through the city to the newly reconstructed Manila (Azcarraga) *matadero* for slaughter. Others, including carabao for work, were released to dealers. In the view of the authorities, this facility promised some important protection from the transmission of bovine disease—much better protection than the private corrals of the dealers scattered about the city. The facility at Pandacan became the place where animals from Cambodia and Vietnam were landed, quarantined, and inspected. After some time, animals were slaughtered there as well. A similar facility was set up and maintained at Iloilo, but not at Cebu, which was now effectively closed to live-animal imports.

Cattle continued to arrive from the ports of China during the first months of 1911 but were held on lighters in the bay for ten days before being certified as disease free and allowed to land. Not surprisingly, given the history of animals coming from these sources, both rinderpest and hoof-and-mouth disease appeared among the stock thus quarantined. The Lichaucos report, without definite date, that disease once forced Faustino Lichauco to dump an entire shipload of cattle into Manila Bay. This may have been that time. In any case, the authorities feared that the workers tending the animals on the lighters would spread the disease in Manila and thence to the country at large—much as cattle-coolies had spread the disease inadvertently among several separate dairies in Shanghai. Accordingly, after a decade of battling the importers, the flow from China was effectively stopped by imposing an uneconomic three-month-quarantine requirement. Shipments from China were quickly replaced by major arrivals from Australia, specifically from the little port of Wyndham on the north coast of Western Australia. The importers of Australian animals were required to build holding pens and an abattoir across the bay thirty miles from the city at a place called Sisiman on the Bataan Coast. Wild animals and local domestic stock were kept well away from any potential disease contact by effective fencing. Starting in 1911, the Australian arrivals in their thousands were landed and slaughtered at Sisiman with the sides of beef delivered to Manila daily by steamer. The new quarantine system was not ideal—the Bureau of Agriculture would have preferred to meet the demand in the short-to-medium term with imports of chilled beef rather than live animals, and there was a continuing risk from the diseased animals that appeared in the city at the Pandacan facility—but it was better than the amoral chaos of prior practice.[29]

The new quarantine and quick slaughter system at Sisiman worked effectively to stop the real threat of bovine-disease introductions from Australia, but the stations at Pandacan and Iloilo both released animals to dealers for sale and were not immediately successful at stopping the further introduction of disease. Rinderpest and hoof-and-mouth disease were discovered in animals under quarantine, and outbreaks of rinderpest in Laguna and Rizal provinces near Manila and in Iloilo and Capiz provinces in 1912 were all traced to cattle recently arrived from Indochina. "These animals had been passed by a veterinarian of the [Philippine] Bureau of Agriculture in Indochina, by a French veterinarian there, and besides had undergone ten days quarantine in the Philippines," reported one source.[30] In their defense, resistant animals and those at the end of a period of immunity may harbor a mild and not-easily-detectable but readily transmittable form of the disease. In any case, as a result of these new outbreaks, the ninety-day-quarantine rule was extended to animals arriving from this region, and for the next two years, few cattle were imported from Indochina. In 1912, cattle importers were offered simultaneous inoculation of their stock in Hong Kong or Phnom Penh at their own expense as an alternative to three months' quarantine, but this was not adopted because of the high death rates following the application.

The hiatus in continual reinfection from abroad together with the provincial quarantine efforts in Pangasinan and elsewhere were surely the major reasons that the period from mid-1911 through 1915 stands as the low point of infections and bovine deaths between the second and third waves of the Philippine rinderpest epizootic. The other reason for the lower death rate was that rinderpest was becoming enzootic in some places like Pangasinan, Panay, and the provinces around Manila. The mortality in areas of chronic outbreaks could be much less than elsewhere, even as low as 20 percent. There were now only a few pockets of animals that had not already been exposed; the best known of these were on small islands. The same period saw considerable recovery in domestic livestock numbers.[31]

The Third Wave

Despite the clear public interest in controlling animal disease, bovine imports from Indochina and China soon resumed in earnest, amounting in

1915 to perhaps sixteen thousand head. Prices of meat had risen by 100 percent in the last few months of 1913 due to the shortage of domestic cattle. Agitation by affluent traders and beefeaters proceeded apace. Predictably, the third-wave rinderpest epizootic exploded early in 1916, affecting eighteen provinces by the end of the year. Because of more effective intervention and perhaps because of the effect of more regular disease exposure, the prolonged third wave failed to develop the same intensity as the first two, with annual mortality peaking in 1917 at twenty-seven thousand and again in 1921 and 1922 at thirty-five thousand. It took eleven years to reduce the annual bovine mortality from this disease below ten thousand, but in this entire period, fewer than half as many animals died as during 1902 alone in the second wave.[32]

The Veterinary Division of the Bureau of Agriculture attempted to respond vigorously to the new epizootic, but at first with resources and biological weapons that were grossly unequal to the task. Medical interventions that worked moderately well in India and Shanghai on more resistant animals either did not work or caused high mortality in the Philippines. A major change from the two earlier waves, however, is that there now existed an immunization that was not lethal to animals in good condition and that was often successful in conferring a long-lasting immunity. Many animals arriving from abroad, however, were not in good condition. Despite the bureau's efforts, during 1918, the third year of the outbreak, only three provinces in northern Luzon had been cleared (for the time being). The epizootic continued active in twenty-seven other provinces, plus Manila, and had newly spread to five more from Davao to Bikol to Ilocos Norte. The largest numbers of deaths that year were recorded in the inner zone around the city.

The bureau attempted a major campaign on Masbate Island during 1918. It was clear that "smugglers" pursuing their private interests were avoiding the restrictions on movement. The bureau suspected that this particular outbreak occurred when a local dealer tried to carry a few cattle to Leyte for sale; unable to make a profit, he returned the animals to Masbate. In the meantime, they had contracted the disease. The bureau responded with a handful of veterinarians, thirty livestock inspectors, and fifty constabulary troops as quarantine guards. On this occasion, the disease was stopped just beyond the municipality of the first outbreak—no mean feat on an island with an open-range cattle economy and few fences. In 1921, on the front of the second peak in this wave, the greatest mortality was now recorded in the Western and Central Visayas. In these peak years of the third wave, the bureau calculated that the annual death rates were just less than 3 percent, a far cry from the devastation of the first and second iterations. Finally, in 1925, the disease again invaded the Ilocos coast and the Mountain Province in the north. Starting at the southern tip of La Union and running northward in a chain of infection among animals grazing on the hillsides, it was finally stopped at Tagudin in southern Ilocos Sur by targeting a mass vaccination campaign just ahead of the disease. Again, the Philippine Constabulary, three hundred strong, established an effective quarantine cordon, running from the coast to the hills. Owners of semi-wild cattle grazing in the hills were warned that animals found running loose would be shot. In addition to these efforts, an effective vaccine was now available and was aggressively given in mass campaigns to from two hundred thousand to three hundred thousand animals a year from 1924 through 1931. By 1927, losses from this disease were under three thousand per year.[33]

An effective vaccine suitable to Philippine conditions was developed in stages by the scientists of the Bureau of Agriculture. In 1923, William H. Boynton, a pathologist with the bureau, developed a tissue vaccine incorporating finely ground material from the organs of infected animals. In careful application by well-trained personnel, it represented a breakthrough that helped greatly to lower rinderpest mortality and eradicate the disease from some enzootic areas. In 1927, a similar vaccine was treated with chloroform. This could be more readily prepared and kept in refrigeration for extended periods. A practical drawback to both vaccines was that they required three injections over a period of weeks. Finally, in 1934, the Filipino veterinarians M. M. Robles and J. D. Generoso developed a dried vaccine that could be kept a month at room temperature and more than two years under refrigeration. Further, it required only a single injection. With an end to regular imports of live bovines and the widespread application of the improved vaccines, the incidence of rinderpest declined rapidly. Progress against the disease was now such that "scouting parties composed of veterinarians and livestock inspectors" could be sent to scour the outbacks for hidden cases. The last case was found in the wilds of southern Negros in 1938. After half a century of intermittent devastation, the combination of vaccine, quarantine, and near-zero imports worked—a very substantial public health and economic achievement. The eradication of rinderpest in the Philippines has proved long lasting.[34]

After heavy and prolonged lobbying, the politics of the Great Depression put an end to imports of live cattle for slaughter in mid-1930 and for the rest of the pre–World War II era. It was in this protectionist environment that ranches in Bukidnon Province, specializing in hybrid zebu animals, rose to substantial financial success.

MANILA LONG remained the principal market for beef cattle, with relatively few being consumed in the provinces. Growth in urban demand, more productive competing uses for the land, an ongoing technological transition in shipping and port infrastructure, crossbreeding with imported Indian varieties sponsored by the agricultural bureaucracy, and legislation aimed at disease control and protectionism each played a role in shaping the directions of change in the Philippine bovine industry. Increasing participation in interregional and global trade led to the diffusion of rinderpest to the Manila and Iloilo areas as it did to Java, Sumatra, and Sulawesi in the Indonesian archipelago at roughly the same time.

The first two waves of rinderpest in the Philippines appear like biblical plagues. They completely devastated lowland wet-rice production and immediately changed the pattern and volume of flow of beef animals to the city. In each case, the numbers of carabao and cattle were greatly reduced, but the reductions were not proportional. Between 1870 and 1903, epizootics reduced the gross numbers of carabao by 40 percent and the number of cattle by 77 percent—crediting the numbers recorded in both cases. This disproportionality changed the ratio between the two species from fifty-one cattle per hundred water buffalo to only twenty. Clearly, carabao were needed for muddy-field preparation and even for transportation of produce into the city on the quagmires that passed for roads during the rainy seasons in the nineteenth century and early twentieth. Cattle had a broad range of useful characteristics including the ability to work on hard surfaces and for greater periods in high heat, but from a Manila perspective, it was not critical that they be raised locally. For use as meat, they could be imported. So, the number of carabao in the country recovered first, despite their slower reproduction rate. Even on the eve of World War II, the ratio between these two animals stood at forty-six cattle per hundred carabao— not quite back to the level of 1870—and this despite the large numbers of cattle now being raised for beef. This was not universally the case in insular Southeast Asia following the great waves of rinderpest, but the Philippines had a special dedication to the carabao.[35]

It happens that I live in a dairy state, so bovines are important. Dairymen and other domestic stock raisers avoid bringing cows into our state from nearby provinces without the proper veterinary examination and documentation. If someone takes a shortcut and a disease such as brucellosis breaks out, his herd is immediately put down, and he is ostracized for having endangered the animals and livelihoods of everyone else. But this is true now in a place with relatively high modern education. In the 1880s and 1900s, many Filipino farmers and stockmen did not have a disease-specific understanding. The Spanish colonial authorities in the 1880s were used to using quarantines for preventing the transmission of human disease from abroad but were not aware of the threat posed by rinderpest. At the same time, the animal science bureaucracy lacked either an effective immunization or treatment. Finally, the combination of biological and epidemiological intervention minimized and then stopped the third rinderpest wave—a formidable achievement, an undisputed good.

Notes

1. In other essays, I have reflected on the evolution of several regional beef-production systems in the Philippines and the critical impact of rinderpest on rice production, urban provisionment, and the balance of foreign trade. "Beef Consumption and Regional Cattle Husbandry Systems in the Philippines, 1850–1940," in *Smallholders and Stockbreeders: Histories of Foodcrop and Livestock Farming in Southeast Asia,* ed. Peter Boomgaard and David Henley (Leiden: KITLV Press, 2004), 307–24; and "Droughts, Rinderpest, and the Rice Deficit," in *Feeding Manila in Peace and War, 1850–1945,* manuscript in progress.

2. This chapter has benefited from critical readings by Matthew Turner and William G. Clarence-Smith. Its shortcomings are my own, alone.

3. John Leake, "The Livestock Industry," *Bulletin of Indonesian Economic Studies* 16, no. 1 (1980): 65. On water buffalo and Bali cattle, see Colin P. Groves, "Domesticated and Commensual Mammals of Austronesia and their Histories," in *The Austronesians: Historical and Comparative Perspectives,* ed. Peter Bellwood, James J. Fox, and Darrell Tryon (Canberra: Australian National University, Department of Anthropology, 1995), 152–63.

4. The rinderpest virus is classed as a morbillivirus and is closely related to viruses causing measles in humans and distemper in canines. *The Merck Veterinary Manual* (Whitehouse Station, NJ: Merck, 2005), 619.

5. *Hungerford's Diseases of Livestock* (Sydney: McGraw-Hill, 9th ed., 1990), 386–87; Geoffrey P. West, ed., *Black's Veterinary Dictionary* (London: A. C. Black 1988), 124–25. See also Vicente Ferriols, J. D. Generoso, and A. B. Coronel, "Rinderpest in the Philippines," *Philippine Journal of Animal Industry* (hereafter *PJAI*) 10, no.3 (1950): 289–306; H. E. Keylock, "The Control of Rinderpest in a Large Dairy Herd in Shanghai," *Journal of Comparative Pathology and Therapeutics* 46, no. 3 (September 1933): 149–58; idem, "Cattle-Plague in China," *Journal of Comparative Pathology and Therapeutics* 22, no. 3 (1909): 193–213; S. Anderson, "An Outbreak of Cattle Plague in China," *Indian Medical Gazette*

36 (September 1901): 327–28; Harold D. Brown, "Rinderpest," *Peking Natural History Bulletin* 4, no. 2 (1929): 87–94.

6. Quotation from H. Warington Smyth, *Five Years in Siam from 1891 to 1896* (London: John Murray, 1898), vol. 2, appendix 6, 285; C. W. Daniels, "The Outbreaks of Rinderpest in Selangor, 1903 and 1904," *Journal of Tropical Veterinary Medicine* 2 (1907): 159–62; and Martine Barwegen, personal communication, 10 March 2003 and chapter 5 of this collection.

7. On variable resistance, see Daniels, "Outbreaks of Rinderpest in Selangor," 159–62.

8. Going further back, the elite of colonial Mexico City routinely ate mutton "stewed or pit roasted, often with chilies in marinade (*adobo*)," and this taste may have carried over among some in Hispanic Manila. Roger Horowitz, Jeffrey M. Pilcher, and Sydney Watts, "Meat for the Multitudes: Market Culture in Paris, New York City, and Mexico City over the Long Nineteenth Century," *American Historical Review* 109, no. 4 (2004): 1066.

9. See the annual *Balanza Mercantil del Comercio de las Islas Filipinas* for the years indicated, as well as Frederick H. Sawyer, *The Inhabitants of the Philippines* (London: Sampsom Low, Marston and Co., 1900), 220; and Fedor Jagor, *Travels in the Philippines* (1873; Manila: Filipiniana Book Guild, 1965), 112, 208.

10. *Merck Veterinary Manual*, 619–20.

11. See Barwegen, chap. 5 of this volume.

12. See Hünniger, chap. 4 of this volume.

13. J. Ph. Hens, Netherlands, *Consulaire verslagen en berichten* (hereafter *CVB*), 1888 (concerning 1887), 1049. My thanks to Chantal Oudkerk Pool for the translation of material from the *CVB*.

14. Gines Geis Gotzens, *Una epizootic en Filipinas* (Manila: Chofre y Compania, 1888), 16–17; Arthur Stanley, "Notes on an Outbreak of Cattle-Plague in Shanghai," *Journal of Hygiene* 2, no. 1 (1902): 43; Jose Vicente Braganza, "Alaminos," *Ilocos Review* 10 (1979): 95–96. The data on cattle shipments are from "Matanza de reses," *El Comercio*, 1 February 1886; 15 January 1887; and 9 January 1888. The Pangasinan numbers are 4,518 in 1885, 3,062 in 1886, and 1,841 in 1887. On the beef supply shortfall in 1888, see "Matanza," *El Comercio*, 11 July 1888.

15. Hens, *CVB*, 1888 (concerning 1887), 1049.

16. "La epizootica," *El Comercio*, 11 January 1891. Vicente Ferriols, "A Brief Resume of Rinderpest Control Work in the Philippines," *Philippines Journal of Agriculture* 1, no. 4 (1930): 393. Ferriols, a veterinary, was chief of the Animal Diseases Control Division of the new Bureau of Animal Industry in 1930.

17. Hens, *CVB*, 1888 (concerning 1887), 1049; and Angel K. Gomez, "Eradication and Control of Rinderpest in the Philippine," *Journal of the American Veterinary Medical Association* 113, no. 857 (1948): 113.

18. John Foreman, *The Philippine Islands* (New York: Charles Scribner's Sons, 1899), 391; Ken De Bevoise, *Agents of the Apocalypse: Epidemic Disease in the Colonial Philippines* (Princeton, NJ: Princeton University Press, 1995), 159–60; Paul Rodell, personal communications, 21 October 1999, and 27 April 2000, including copies of Philippine National Archive documents: Estadistica, Zambales, Ganados, 1886, and "Memorias," Zambales, 1892; Braganza, "Alaminos," 96; and D. F. Doeppers, "A Century of Rice Deficits: El Niños and Rinderpest," in *Feeding Manila in Peace and War, 1850–1945,* ms. in progress. Braganza's numbers using local sources for Alaminos alone indicate a more severe impact than do those for Zambales as a whole where carabao numbers were said to have declined from 11,400 in 1886 to perhaps 8,400 in 1892.

19. *Merck Veterinary Manual,* 620.

20. Stanton Youngberg, "The North to South Movement of Animals on the Island of Luzon," *Philippines Agricultural Review* (hereafter *PAgR*) 5, no. 12 (1912): 654. See also Pablo Tecson, "Agricultural Conditions in Tarlac Province," *PAgR* 1, no. 7 (1908): 301; "Provincial Reports," *PAgR* 3, no. 10 (1910): 589; and Max L. Tornow, "Economic Conditions of the Philippines," *National Geographic Magazine* 10, no. 2 (1899): 40.

21. Norman Owen, *Prosperity without Progress: Manila Hemp and Material Life in the Colonial Philippines* (Berkeley: University of California Press; and Quezon City: Ateneo de Manila University Press, 1984), 181; De Bevoise, *Agents,* chap. 6, esp. 161–63 and 235–37; and *Census of 1903,* vol. 4, 236 and 373–76.

22. Gabriel Alba, acting governor, "Zambales," *Report of the Philippine Commission* (hereafter *RPC*), 1904, pt. 1, 670–71.

23. United Kingdom, *House of Commons Papers,* 1899, vol. 101, 2319, "Trade and Commerce of the Philippine Islands, 1898," 3; United States, Department of State, *Commercial Relations of the United States with Foreign Countries during the Year 1907* (Washington, DC: Government Printing Office, 1908), 1:349; "Zafiro's Big List," *Manila Times,* 14 January 1908; and "Philippine Trade," *Manila Times,* 24 February 1908, 11.

24. "Cattle Had Rinderpest," *Manila Times,* 22 July 1907, 15.

25. Luisa Fernandez Lichauco, *Family Recollections* (Manila: privately printed, 1991), 17, 34; "Cattle Had Rinderpest"; Municipal Board Notes, *Manila Times,* 22 June 1907, and 16 December 1915.

26. Youngberg, "North to South Movement," 654.

27. Ibid., 653–59; *RPC* 1911, 170–71.

28. On the very real threat, see "Cattle Had Rinderpest," *Manila Times,* 22 July 1907, 15; "Cattle Are Infected," *Manila Times,* 30 September 1908; and "Long-Drawn Battle against Rinderpest," *Philippines Free Press,* 12 September 1908, 3.

29. Annual Report of the Bureau of Agriculture (ARBAg), 1910–11, *PAgR* 5, no. 1 (1912): 18–19; and idem, 1911–12, *PAgR* 5, no. 12 (912), xx. See also *Lichauco Family Reunion, 1991* (n.p.: privately printed, 1991), 23; and "Veterinary Work of the Bureau of Agriculture," *PAgR* 1, no. 11 (1908): 447–49. The first law banning the importation of diseased cattle took effect on 1 January 1907. It had little immediate impact. On cattle-coolies, see Stanley, "Notes on an Outbreak," 44.

30. ARBAg, 1911–12, *PAgR* 5, no. 12 (1912): xviii–xix; George S. Baker, "Cattle Importation," *PAgR* 5, no. 12 (1912): 660–61.

31. On lower death rates, see *Black's Veterinary Dictionary,* 16th ed. (1988), 125; and Ferriols, Generoso, and Coronel, "Rinderpest in the Philippines," 291. The government's calculation of carabao numbers reached one million in 1913 for the first time since 1902: "Hits Million Mark Again," *Philippines Free Press,* 14 June, 1913, 2–3.

32. ARBAg, 1918, 118; and *Statistical Bulletin of the Philippine Islands,* no. 11 (Manila: Bureau of Commerce and Industry, 1928), 58. On the commercial pressure to restart live-cattle imports from Hong Kong, see United States, *Daily Consular and Trade Reports,* vol. 17, no. 23, 28 January 1914, 357; no. 38, 14 February 1914, 605; no. 125, 28 May 1914, 1166; as well as vol. 18, no. 231, 2 October 1915, 30; and no. 241, 14 October 1915, 201.

33. ARBAg, 1916, 40–47; 1917, 35; 1918, 108–10; 1921, 26; and 1933, 135. See also *PJAI* 1, no. 6 (Nov.–Dec. 1934), 468; and Antonio Peña, "Cattle Raising in the Philippines," *Commerce and Industry Journal* 6, no. 7 (1930): 7.

34. Ferriols, Generoso, and Coronel, "Rinderpest in the Philippines," 303. For celebration of the Philippine chloroform vaccine abroad, see Brown, "Rinderpest."

35. David Henley, historical agriculture seminar, EuroSEAS Conference, London, 7 September 2001.

CHAPTER 7

Diseases of Equids in Southeast Asia, c. 1800–c. 1945

Apocalypse or Progress?

William G. Clarence-Smith

DESPITE A proliferation of machines, the nineteenth century witnessed the golden age of the horse and the mule in the West, a phenomenon that was replicated in Asia with some variations. As in the West, equids were crucial to military power, urban transport, and elite ceremonies and sports, while playing a limited role in diet outside Central Asia. Equids were thus widely traded by sea and land. Unlike in the West, equids featured little in agriculture and forestry and were used more for pack than for draft in rural transport. However, reliance on equids persisted longer than in the West, where the harsh realities of World War I ensured the unequivocal triumph of the internal-combustion engine.[1]

It is surprising that historians of Southeast Asia have afforded so little attention to equids or indeed to any domestic animals, despite some recent progress.[2] Underlying this neglect is the rarely questioned assumption that tropical diseases prevented the rearing of equids. Traditionally thought of as lying between India and China, Southeast Asia is better pictured as sandwiched between Tibet and Australia. Conditions typical of these two great pastoral zones penetrate deeply into parts of Southeast Asia, notably in higher and drier areas, and relatively low human-population densities also favor the raising of animals.[3]

In the 1930s, Southeast Asia was estimated to contain about 1,750,000 equids. Nearly all were horses, with mules and donkeys generally restricted to the confines of China and Tibet. Mainland Southeast Asia accounted for some 650,000, chiefly bred on high plateaus and in rain-shadow plains.[4] Maritime Southeast Asia contained about another 1,100,000, most intensively raised in the relatively arid Lesser Sunda Islands. Java, northern Sumatra, and southern Luzon were other significant breeding centers.[5]

The need to keep all these beasts alive and working was a significant concern for Southeast Asia's rulers. Indeed, equids were a strategic commodity, given their crucial role in warfare and police duties, so that veterinary medicine emerged in the nineteenth century with a strong initial emphasis on military animals.[6] Local breeds were tough little ponies, with considerable acquired resistance to prevailing ailments and an ability to thrive on local fodder, contrasting with larger and more expensive imported beasts.[7]

Modern methods of combating equine diseases also played an equivocal political role. Initial hecatombs were unleashed on animal populations by campaigns of "pacification," but colonial rulers then introduced novel veterinary structures and methods as part of a wider package of "scientific progress." That said, the racially discriminatory organization of colonial veterinary services undermined claims to legitimacy that flowed from improvements in animal health.

The Curse of the Tropics

Of diseases specific to the tropics, none caused greater problems for equids than *Trypanosoma evansi,* usually known by its Indian name of surra. Provoking serious anemia in equids and camels and usually fatal if untreated, surra is less of a problem for bovids than for equids and does not affect humans. Griffith Evans discovered the trypanosome causing surra in the Punjab in 1880.[8] India's Imperial Bacteriological Laboratory then investigated surra extensively, as did Alexandre Yersin in Vietnam.[9] Evans failed to explain transmission, but research on African trypanosomes directed attention to biting flies, and Rogers first scientifically described infection by tabanids in 1901.[10] Reported across Asia, northern Africa, and Central and South America, surra is caused by protozoan blood parasites, transmitted mechanically by biting flies. As the parasites are almost identical to those causing *Trypanosoma brucei* in sub-Saharan Africa, parasitologists assume that the one evolved from the other "in the last few thousand years," in

Sahelo-Sudanic environments devoid of tsetse flies.[11] They further assume that surra crossed the Sahara with infected camels, probably in the first millennium CE, with Morocco as the focus of further diffusion to Asia, since the parasite tends to get longer the further east it is found.[12] The numerous names for the malady in India, and the resistance developed by Indian cattle, indicate that it has been present for centuries.[13] Surra probably reached China almost as quickly along the various silk roads, although the disease has been poorly investigated in an East Asian context.[14]

In the case of Southeast Asia, Tony Luckins views surra as part of an ecological catastrophe unleashed by Western imperialism in the late nineteenth century.[15] A parallel is drawn with the well-known case of the introduction of Indian cattle infected with surra into Mauritius in 1901, which nearly wiped out the island's numerous equids within a year.[16] However, the crucial assumption that the disease was absent from Southeast Asia before the 1890s needs to be explored.

As far as mainland Southeast Asia is concerned, surra was probably present long before the 1890s, even if colonial campaigns provoked abnormal spurts of the disease. Once established in India and China, there was little to prevent the further spread of surra. Mule and pony caravans linked Tibet and Yunnan to northern Southeast Asia from at least the sixteenth century, and probably from a much earlier date, and they brought horses and mules for sale. Moreover, Chinese emperors sent horses to rulers as gifts.[17]

Surra was certainly described as enzootic in Burma, Thailand, and Vietnam by the early decades of the twentieth century.[18] Malaya's short land frontier with Thailand may have made it part of this same zone, although the disease may also have spread by animals imported by sea.[19]

As for insular Southeast Asia, equids arrived by sea from an early date, for a trade in horses probably linked Bengali ports to the Straits of Melaka in the third century CE.[20] Indian and Chinese horses reached Java from the seventh century.[21] By early modern times, fine horses came from India, and more rarely from the Middle East, Europe, or the Americas. In return, cheap Southeast Asian ponies, especially from Sumatra, Luzon, and Timor, went to eastern India and southern China.[22]

Dutch reports on equine mortality in early nineteenth-century Indonesia are hard to interpret. Some fatal sickness afflicted Central-Java's horses in 1819, also affecting cattle.[23] The British may thus have brought surra to Java with the Indian horses that they employed for the conquest and occupation of the island from 1811 to 1816.[24] Southeastern Borneo suffered outbreaks of an "incomprehensible" disease in 1830–31, 1839, and 1842, suggesting that this was not a malady known in Europe.[25] Both North and South Sulawesi lost numerous horses to unidentified sicknesses in the 1840s and 1850s.[26]

Dutch veterinarians first unambiguously diagnosed surra in Javanese horses in 1897, attributing mortality to the disease back to 1886.[27] It may have contributed to a marked decline in the horse population of Java and Madura between 1880 and 1900, amounting to some one hundred thousand head, or 20 percent of the total, with the main reduction occurring between 1895 and 1900.[28] However, this can also be explained by a contraction in pasture, resulting from rapidly expanding population, which led to a greater reliance on ponies imported from the Lesser Sundas.[29] Surra was considered enzootic in the Javanese lowlands by 1907, which suggests a longer presence on the island.[30] In addition, the Javanese were already well aware of the transmission of parasites by biting flies, which the Dutch refused to believe until the turn of the century, while greater resistance in Madurese than in Javanese cattle points to regional diversity.[31] Some evidence of trypanotolerance was noted in Lombok horses in 1996, but it is unclear when this might date from.[32]

The case of West Timor is ambiguous. From around 1912, the Dutch introduced Balinese cattle, bringing with them new *Hippobosca* bloodsucking flies. These insects soon became a pest at lower altitudes, forcing horse breeding to retreat into the highlands.[33] Swarms of *Hippobosca* were later accused of transmitting surra in Portuguese East Timor.[34] However, it is far from clear that surra was absent in West Timor at the time, as an undefined epizootic, culminating in 1891, killed many horses.[35] This was followed by the death of some 80 percent of the numerous ponies on the offshore islands of Roti and Savu in 1905.[36]

The Philippines provide the most convincing case of an apocalyptic visitation under high colonialism. A long-time resident in the colony declared that surra was "unknown in these islands before the American advent," whereas it was "common in British India." As American troops conquered the archipelago from 1898, the disease spread like wildfire, provoking mortality rates among ponies and mules as high as 60 percent.[37] Together with other diseases, surra may have killed "as much as 80 percent" of the stock of horses by 1908.[38] That said, the impact of surra was worse among American and Australian equids, suggesting a degree of tolerance in locally bred ponies.[39]

Once the transmission of surra was properly understood, veterinarians concentrated on insect control, quarantine, and culling. Scientists in Kuala Lumpur experimented with excluding flies from stables from 1901, arguing that only *Tabanus* flies, found in the open, spread the parasites.[40]

However, the Dutch also blamed *Stomoxys* flies, typically encountered in stables. They drove the insects away with smoke and stabled animals during the day.[41] Attacking insects was of dubious efficacy, however, in view of the numerous and ubiquitous potential vectors.[42] As blood tests determined infestation, the focus moved to quarantine and culling, but without eradicating the disease.[43] At best, this strategy prevented surra from taking hold in new areas, as in northwestern Australia in 1907 when infected camels were detected.[44]

Chemical treatments thus moved to the fore, with advice on "different arsenical preparations" published in India from 1906.[45] The major colonial powers experimented with arsenic and antimony in the interwar years, and Naganol emerged as the best drug.[46] It remained the mainstay for decades, but it was expensive and could not entirely eradicate surra. It is a concern that resistance was reported in Vietnam in the 1990s.[47]

Easily confused with surra is equine piroplasmosis, also known as tick plague, biliary fever, or Texas fever, to which horses are somewhat more susceptible than donkeys or mules. Two similar protozoan blood parasites, *Babesia equi* and *Babesia caballi*, destroy red blood cells. These protozoa are transmitted by many blood-sucking ticks, within which parasites reproduce sexually.[48] Piroplasmosis was among the major enzootic equine diseases of Indochina and Indonesia, with local horses demonstrating some resistance.[49] The Americans used an unspecified "biologic" against the malady in the Philippines, with some success.[50]

Global Diseases

Many equine diseases afflicting Southeast Asia were not specifically tropical, notably glanders, the main nineteenth-century global threat to horses, and research on these maladies was carried out mainly in the West.[51] Glanders was even more dangerous to donkeys and mules than to horses and occasionally passed to humans in intimate contact with equids. Caused by the bacterium *Burkholderia mallei* (formerly *Pseudomonas mallei*), glanders is transmitted between equids by contaminated food and water, inhalation, or contact. Ulcerations can be pulmonary, nasal, or cutaneous. The pulmonary form is the most dangerous, while the cutaneous kind was long thought to be a separate disease, farcy. European scientists identified the bacterium in 1882, failing to develop a vaccine but producing mallein as a valuable diagnostic test from 1891. As glanders has no wild hosts, culling infected animals proved highly effective. The mallein test further distinguished between glanders and epizootic lymphangitis, caused by *Burkholderia pseudomallei*, which presents similar symptoms but affects mammals other than equids.[52]

Both glanders (Dutch *kwade droes*) and epizootic lymphangitis were probably ancient diseases in Southeast Asia. Infection rates for glanders were highest where numerous equids were in close contact, notably in coastal towns and on caravan routes.[53] Mortality from glanders was high during Spanish and American military campaigns in the Philippines, but the major islands were already familiar with the disease.[54] However, some Southeast Asian islands were spared until the spread of steam navigation in the late nineteenth century.[55]

As the mallein test came to be produced locally and glanders became a legally notifiable sickness, culling infected animals and burning the carcasses became the chief method of control.[56] The Dutch thus almost eliminated the disease in the island of Lombok after 1919 but encountered stiff indigenous opposition to testing further east. They thus concentrated on testing ponies imported into Java from the Lesser Sundas.[57] Financial cuts during the Great Depression led to the cancellation of compensation to horse owners in 1932, causing the disease to flare up again in certain Burmese towns.[58]

Strangles (Dutch *droes*) was again specific to equids around the globe, but it was much more benign than glanders. The bacterium *Streptococcus equi* caused abscesses in lymph nodes beneath and behind the jaw, compressing the pharynx and making breathing difficult. Once abscesses burst, most horses recovered quickly and completely, despite occasional complications such as pneumonia. Like mumps in humans, one attack usually conferred immunity for life.[59] Strangles occurred all over Southeast Asia, typically among young horses.[60]

Anthrax (French *charbon*; Dutch *miltvuur*) afflicts all warm-blooded animals, including humans. Septicemia results from infection, leading to rapid death. As infected carcasses release long-lived spores of the bacterium *Bacillus anthracis*, the surest preventative measures are cremation or deep burial with quick lime. Following Louis Pasteur's discoveries, vaccination proved effective against the disease.[61] Anthrax was endemic in Southeast Asia and periodically inflicted high mortality on horses, despite their lesser susceptibility than cattle.[62] Cost limited the availability of the vaccine, but it proved its worth in interwar Burma and the Philippines.[63]

Tetanus, or lockjaw, was another global bacterial affliction of mammals. The anaerobic *Clostridium tetani*, ubiquitous in soil, develops in damaged tissues, and the toxins cause muscular spasms, paralysis, and often death. For equids, puncture wounds in hooves are the most common entry points. A vaccine and serum were developed but were costly.[64] Tetanus thus remained among the common causes of equine mortality in Southeast Asia.[65]

The Development of Veterinary and Research Structures

The delivery of services to animals became more specialized, as the scientific revolution gathered pace in the nineteenth century. Modern states in Southeast Asia, mainly colonial but including independent Thailand, initially appointed veterinarians to look after military animals. It was only slowly that officials set up civilian veterinary services and even more slowly that research became separate from treatment.

Research and publication remained generally free from nationalistic rivalries and received a stimulus from two outside bodies. One was the network of Instituts Pasteur, radiating out from Paris from 1888, researching both human and animal diseases. Some Instituts Pasteur were located in French Indochina, but three were founded in India between 1900 and 1917, one in Bangkok in 1912, and one in Rangoon in 1915.[66] The other outside stimulus came from the Imperial Bacteriological Laboratory of British India, renamed Imperial Institute of Veterinary Research in 1925. Founded in Poona in 1889, it moved to Mukteswar, west of Nepal, in 1893, with an offshoot at Izatnagar in the North Indian plains from 1913.[67]

Burma, part of British India till 1937, had its own Veterinary Department from 1874. After the conquest of Upper Burma in 1885–86, veterinarians were charged with checking diseases brought by overland caravans. Urban ponies were the chief patients in a veterinary hospital opened in Insein, near Rangoon, in 1911, and the busy laboratory was boosted by the arrival of a research officer from South Africa in 1927. Dispensaries appeared in several other towns after 1921. The department's staff temporarily contracted in the recession of the early 1930s, but by 1939 there was a veterinarian in each of the colony's forty districts.[68] At this stage, their chief efforts went into inoculation.[69] Autonomous Shan princes in the highlands hired their own veterinary assistants from around 1900, reducing the incidence of disease by quarantining sick ponies and mules.[70]

Quarantine loomed large in Malaya, for the peninsula bred virtually no equids, depending on imports.[71] Private Western veterinarians were in Singapore from 1860, a municipal service emerged in the 1920s, and government departments slowly emerged in the protected Malay sultanates. Animal depots, later infirmaries, treated animal outpatients from 1904. Kuala Lumpur's Pathological Institute opened in 1901, under an American director. It researched both human and animal ailments and was affiliated with the London School of Tropical Medicine.[72]

Dutch veterinary surgeons worked in military studs from 1814, and there were a few private and municipal veterinarians from early in the century. A civilian service emerged in 1853, limited to Java. It expanded from the 1880s, responding to a wave of epizootics, but contracted again in the economic recession of the early 1930s. Research on animal maladies took off in the 1890s, and a specialized veterinary laboratory emerged in Buitenzorg [Bogor] in 1907.[73]

French Indochina's army studs, concentrated in northern Tonkin close to the sensitive Chinese frontier, initially monopolized veterinarian work, but a civilian Service Vétérinaire et Zootechnique later emerged, charged with checking animal diseases in trading caravans.[74] Research was concentrated in Annam's Nha Trang Pasteur Institute, founded in 1895 and directed by Alexandre Yersin, the famous Franco-Swiss doctor who identified the plague bacillus in Hong Kong in 1894. He raised animals, including horses, to conduct experiments.[75]

The American conquest of the Philippines from 1898 resulted in a flurry of activity, although the Spaniards had already created a health inspectorate for livestock in 1849, followed by a civilian veterinary service. The Americans generously funded a new Veterinary Division from 1901, which killed and cremated infected horses imported by sea. A law of 1907 created permanent quarantine stations and imposed fines and imprisonment for the movement of sick animals. An animal clinic in Manila provided free consultation and treatment for urban horses from 1912. Laboratories from 1901 produced serums and vaccines, and a specialized Veterinary Research Division emerged in 1929, in the newly created Bureau of Animal Industry.[76]

Independent Thailand lagged behind somewhat, but the government imposed animal-quarantine measures in 1897.[77] Railway construction stimulated "the fight against contagious diseases of draft animals by veterinarians" in 1900 on the Khorat Plateau, the country's chief pastoral zone.[78] A Veterinarian Division existed in the Ministry of Lands and Agriculture in 1930, in which the number of veterinarian officers was reported to have risen from forty-two to 101.[79]

Southeast Asia's veterinary programs largely collapsed during the conflict of 1941–45, despite Japanese efforts to maintain existing structures. Epizootics raged in Burma and the Philippines.[80] In Cambodia, surra contributed to a drastic fall in horse numbers.[81] Requisitioning, theft, and neglect of insect control reduced the equine population of the Lesser Sundas.[82] Lax quarantine precautions allowed glanders, surra, and epizootic lymphangitis to spread in Malaya.[83] Even India suffered problems with surra, due to a shortage of drugs.[84]

Asian Personnel and Traditional Healers

Racial hierarchies undermined the efficacy of colonial veterinary services, with "Europeans" generally monopolizing top positions. "Foreign Asians," often from India, were in the middle. As for indigenous people, they were few in number and junior in rank. Independent Thailand may have been something of an exception to this generalization, as a 1930 report failed to mention the "race" of veterinary officers.[85]

Of all the colonial powers, the Americans probably made the greatest efforts to promote local veterinarians, albeit with disappointing results. An agricultural college opened in 1909 at Los Baños, close to Manila, and the animal husbandry building later became the College of Veterinary Service.[86] From 1913, a Democrat administration in Washington decided to appoint Filipinos whenever possible, leading to an exodus of American personnel. Young Filipinos were reluctant to study veterinary science, however, with only thirty-four students in the college in 1925. This prompted the caustic comment that "until there is a livelier and more general interest in saving carabao [buffalo] than in painting them, the country will not attain a high degree of material prosperity."[87]

In Burma, the British made considerable efforts to train locals. The duties of veterinary surgeons included educating Burmese assistants, and the Rangoon Veterinary College opened in 1891, albeit only for "second-class" Burmese and Indian assistants, of whom there were 168 in 1920. A new college was built at Insein in 1925–26, for both training and refresher courses, but enrollment remained modest. Indeed, it ceased altogether from 1929 to 1933 because of cut-backs during the Depression.[88]

The British relied more heavily on South Asians in Malaya, drawing on a growing pool of qualified personnel. An All-India Veterinary School was founded in 1877 and moved to Lahore in 1882, and other teaching institutions were dotted around the subcontinent.[89] Shortages of qualified Europeans during World War I led to a surge in recruitment of Indians, often Goans, who remained confined to junior posts. The first Malay veterinarian was appointed only in 1920, and no more was mentioned until 1947.[90] A Penang Chinese researcher, with a medical degree from Cambridge, was employed in 1903, but he was seen as an exceptional figure.[91]

In Indonesia and Indochina, "foreign Asians" were not mentioned, and the "European" legal category was more likely to include assimilated and mixed-race persons. In Indochina in 1930, there were thirty-eight European veterinarians and eighty-seven "native assistants."[92] A Dutch veterinary school, set up in Surabaya in 1860, closed in 1876 because of poor results.

"Natives" then apprenticed themselves to "Europeans," but the Dutch allowed them only to be assistants. Formal education for Indonesians resumed in a new school in Buitenzorg in 1904, and a few went to the Netherlands in the 1920s, while increasing numbers of less-schooled "paravets" found employment in minor roles.[93]

It took the Japanese conquest for non-European veterinary personnel to achieve rapid promotion. As whites fled or were imprisoned, Asians filled their places, albeit with a small number of Japanese veterinarians occupying the highest positions.[94] Nationalist agitation, fanned by the departing Japanese, made it very difficult to reverse these changes after the war.

The dominance of white veterinary officers, with a poor command of indigenous languages and an arrogant attitude toward local knowledge, undermined the impact of Western veterinary medicine and helped to perpetuate Non-Western forms of healing.[95] Moreover, "scientific" veterinary medicine was both expensive and erratic, whereas local healers had an intimate acquaintance with their environment. Indeed, the British grudgingly admitted that Indian horse doctors, despite all their "mumbo-jumbo," were "acquainted with a few valuable remedies."[96]

Under a thin veneer of Western veterinary medicine, there thus survived a flourishing indigenous sector. Malay healers, Chinese herbalists, and Indian Ayurvedic specialists competed long after the arrival of Western-trained personnel in Malaya.[97] "Farriers" and "native practitioners" in the Philippines were popular because they were cheap and made house calls, using bleeding, leeches, and acupuncture. They persisted in their calling, despite a raft of regulations aimed at forcing them out of business.[98]

INDEPENDENCE BROUGHT a growing belief that techniques relating to equids were "backward," relegating knowledge and skills accumulated over centuries to folk memories and dusty volumes. In Indonesia, the very language of modern scientific research became largely unintelligible to new generations, no longer educated in Dutch. Yet, equids represent a valuable niche technology, regaining favor in the West in sectors such as forestry and livestock management. Moreover, sports based on horses have become lucrative businesses, especially racing and polo. It is thus time to seek to understand afresh how best to care for Southeast Asian equids.

Of necessity, the study of disease will be an important component of any such project. Given the fragmentation of Southeast Asia, in terms of states, indigenous and colonial languages, and even scripts, it will be difficult to achieve any convincing overall picture until national and regional histories have been compiled. Abundant primary and secondary materials exist to write such veterinary stories, but little has yet been done to exploit them.

One problem that looms large in writing about the evolution of equine diseases is how to interpret the unscientific accounts of maladies before the late nineteenth century. The dates at which surra and glanders began to afflict the whole region, or parts of it, thus remain uncertain. This makes it particularly difficult to evaluate the effects of new veterinary services in terms of animal welfare.

A related problem is how to interpret the impact of veterinary services on imperialism and nationalism. Old colonial hands emphasized a linear and cumulative "triumph of science," whereas postcolonial writers give much weight to indigenous Asian medical traditions. However, from what has been written to date, it seems that Southeast Asian farmers were intensely practical and eclectic. They resisted most colonial breeding programs because they were poorly planned and executed, producing equids that were vulnerable to prevailing diseases and that rejected local fodder. In contrast, once the benefits of vaccines and drugs had been demonstrated, livestock breeders kept asking for more. As for the racial ordering of veterinary services, it is necessary to go beyond charting its existence, by understanding how it was perceived by Asians working within these new structures.

Notes

1. Marylian Watney and Sanders Watney, *Horse Power* (London: Hamlyn, 1975); Juliet Clutton-Brock, *Horse Power: A History of the Horse and the Donkey in Human Societies* (Cambridge, MA: Harvard University Press, 1992); Alexander T. Yarwood, *Walers: Australian Horses Abroad* (Melbourne: Melbourne University Press, 1989); Greg Bankoff, "A Question of Breeding: Zootechny and Colonial Attitudes towards the Tropical Environment in the Late Nineteenth-Century Philippines," *Journal of Asian Studies* 60, no. 2 (2001): 413–38; William G. Clarence-Smith, "Cape to Siberia: The Indian Ocean and China Sea Trade in Equids," in *Maritime Empires: British Imperial Maritime Trade in the Nineteenth Century*, ed. David Killingray, Margaret Lincoln, and Nigel Rigby (Woodbridge, England: Boydell and Brewer, 2004), 48–67; Greg Bankoff and Sandra Swart, eds., *Breeds of Empire: The Invention of the Horse in Southeast Asia and Southern Africa, 1500–1950* (Copenhagen: NIAS Press, 2007).

2. Peter Boomgaard and David Henley, eds., *Smallholders and Stockbreeders: Histories of Foodcrop and Livestock Farming in Southeast Asia* (Leiden: KITLV Press, 2004), pt. 2, chapters by William Clarence-Smith, Peter Boomgaard, Greg Bankoff, Martine Barwegen, and Dan Doeppers.

3. William G. Clarence-Smith, "Horse Breeding in Mainland Southeast Asia and Its Borderlands," in ibid., 189–210; idem, "Horse Trading: The Economic Role of Arabs in the Lesser Sunda Islands, c. 1800 to c. 1940," in *Transcending Borders: Arabs, Politics, Trade, and Islam in Southeast Asia,* ed. Huub de Jonge and Nico Kaptein (Leiden: KITLV Press, 2002), 143–62.

4. Clarence-Smith, "Horse Breeding," 189.

5. Pierre van der Eng, *Agricultural Growth in Indonesia since 1880* (Groningen: Rijksuniversiteit Groningen, 1993), 279–81; Joachim K. Metzner, *Man and Environment in Eastern Timor: A Geoecological Analysis of the Baucau-Viqueque Area as a Possible Basis for Regional Planning* (Canberra: Australian National University, 1977), 194; Frederick L. Wernstedt and Joseph E. Spencer, *The Philippine Island World: A Physical, Cultural and Regional Geography* (Berkeley: University of California Press, 1967), 649.

6. Martine Barwegen, "Gouden hoorns: De geschiedenis van de veehouderij op Java, 1850–2000" (PhD diss., University of Wageningen, 2005), 67.

7. G. J. Younghusband, *Eighteen Hundred Miles on a Burmese Tat, through Burmah, Siam and the Eastern Shan States* (New Delhi: Asian Educational Services, 1995), 14, 27–28, 76–77, 144.

8. Griffith Evans, "On a Horse Disease in India Known as 'Surra,' Probably due to a Haematozoon," *Veterinary Journal and Annals of Comparative Pathology* 13 (July 1881): 1–10, 82–88, 180–200, 326–33.

9. Sir Frank Ware, "India," in *A History of the Overseas Veterinary Service, Part I,* ed. G. P. West (London: British Veterinary Association, 1961), 25, 32; Philippe Buchy, "L'Institut Pasteur de Nha Trang," *Association Adaly,* http://www.adaly.org/buchy.htm.

10. Cecil A. Hoare, *The Trypanosomes of Mammals: A Zoological Monograph* (Oxford: Blackwell, 1972), 571.

11. Wendy Gibson, "Sex and Evolution in Trypanosomes," *International Journal of Parasitology* 31 (2001): 644.

12. Hoare, *Trypanosomes,* 557, 567–68, 579, 582–86, 593.

13. A. S. Leese, *A Treatise on the One-Humped Camel in Health and in Disease* (Stamford, England: Haynes & Son, 1927), 224; Hoare, *Trypanosomes,* 555, 581.

14. H. G. Fletcher, *Tengyueh* (Shanghai: Maritime Customs, 1927), 98.

15. A. G. Luckins, "*Trypanosoma evansi* in Asia," *Parasitology Today* 4, no. 5 (1988): 138–39.

16. *Institute for Medical Research, 1900–1950* (Kuala Lumpur: Institute for Medical Research, 1951), 306.

17. Clarence-Smith, "Horse Breeding," 189–98.

18. *The Imperial Gazetteer of India,* vol. 22 (Oxford: Clarendon Press, 1908), 258; Fletcher, *Tengyueh,* 98; Executive Committee of the Eighth Congress of the

Far Eastern Association of Tropical Medicine, *Siam in 1930: General and Medical Features* (Bangkok: White Lotus, 2000), 265; Albert Tricard, "Le cheval en Indochine," *Bulletin Économique de l'Indochine* 27, no. 166 (1924): 213.

19. *Institute for Medical Research*, 306–7.

20. Ranabir Chakravarti, "Early Medieval Bengal and the Trade in Horses: A Note," *Journal of Economic and Social History of the Orient* 42, no. 2 (1999): 205.

21. Han Knapen, *Forests of Fortune? The Environmental History of Southeast Borneo, 1600–1880* (Leiden: KITLV Press, 2001), 294; I. H. Burkill, *A Dictionary of the Economic Products of the Malay Peninsula*, vol. 1 (Kuala Lumpur: Ministry of Agriculture and Co-operatives, 1966), 1216.

22. Takeshi Ito, "The World of the Adat Aceh: A Historical Study of the Sultanate of Aceh" (PhD diss., Australian National University, 1984), 376–79, 413; Serafin D. Quiason, *English "Country Trade" with the Philippines, 1644–1765* (Quezon City: University of the Philippines Press, 1966), 67, 73, 94, 105, 172; Roderich Ptak, "Pferde auf See, ein vergessener Aspekt des maritimen chinesischen Handels im frühen 15 Jahrhundert," *Journal of the Economic and Social History of the Orient* 34, no. 2 (1991): 221; and Austin Coates, *China Races* (Hong Kong: Oxford University Press, 1994), 15–16, 29, 63, 72.

23. Elly M. C. van Enk, "Britse kooplieden en de cultures op Java: Harvey Thomson, 1790–1837, en zijn financiers" (PhD diss., Vrije Universiteit Amsterdam, 1999), 160.

24. John Crawfurd, *A Descriptive Dictionary of the Indian Islands and Adjacent Countries* (Kuala Lumpur: Oxford University Press, 1971), 154–55; J. H. Moor, *Notices of the Indian Archipelago and Adjacent Countries* (London: Frank Cass, 1968), 190.

25. Han Knapen, "Epidemics, Droughts and other Uncertainties in Southeast Borneo during the Eighteenth and Nineteenth Centuries," in *Paper Landscapes: Explorations in the Environmental History of Indonesia*, ed. Peter Boomgaard, Freek Colombijn, and David Henley (Leiden: KITLV Press, 1997), 139; Knapen, *Forests of Fortune?* 296.

26. David Henley, *Fertility, Food and Fever: Population, Economy and Environment in North and Central Sulawesi, 1600–1930* (Leiden: KITLV Press, 2005), 14–15; *Koloniaal Verslag 1860* (The Hague: M. Nijhoff, 1860), 146.

27. Barwegen, "Gouden hoorns," 129–30.

28. *Encyclopaedie van Nederlandsch-Indië*, 1st ed., vol. 4 (The Hague: M. Nijhoff, 1896–1905), 504–5.

29. Sibinga Mulder, "De economische beteekenis van het vee in Ned: Oost-Indië en de regieringszorg ervoor," *De Indische Gids* 19, no. 1 (1927): 308–9.

30. T. W. Jones, R. C. Payne, I. P. Sukanto, and S. Partoutomo, "*Trypanosoma evansi* in the Republic of Indonesia," *Proceedings of the First Internet Conference*

on *Salivarian Trypanosomes*, FAO Corporate Documents Repository, http://www.fao.org/docrep/W5781E/w5781e05.htm.

31. Barwegen, "Gouden hoorns," 130, 134.

32. Simon A. Reid, "*Trypanosoma evansi* Control and Containment in Australasia," *Trends in Parasitology* 18, no. 5 (2002): 220.

33. J. Ormeling, *The Timor Problem: A Geographical Interpretation of an Underdeveloped Island* (Groningen: J. B. Wolters, 1956), 122–23, 159, 167–28, 191.

34. Metzner, *Man and Environment*, 199.

35. I Gde Parimartha, "Perdagangan dan politik di Nusa Tenggara, 1815–1915" (PhD diss., Vrije Universiteit Amsterdam, 1995), 216–17.

36. *Encyclopaedie van Nederlandsch-Indië*, 2nd ed., vol. 4 (The Hague: M. Nijhoff, 1917–1921), 523.

37. J. Foreman, *The Philippine Islands* (Shanghai: Kelly and Walsh, 1906), 336, 338, 622.

38. Greg Bankoff, "Horsing Around: The Life and Times of the Horse in the Philippines at the Turn of the Twentieth Century," in Boomgard and Henley, *Smallholders and Stockbreeders*, 241.

39. Charles B. Elliott, *The Philippines to the End of the Commission Government* (New York: Greenwood, 1968), 346.

40. W. E. Lancaster, "Malaya," in West, *History of the Overseas Veterinary Service*, 95, 98, 102; *Institute for Medical Research*, 53, 307–8.

41. *Encyclopaedie*, 2d ed., vol. 4, 523; Jones, Payne, Sukanto, and Partoutomo, "*Trypanosoma evansi*"; Luckins, "*Trypanosoma evansi* in Asia," 140.

42. Barwegen, "Gouden hoorns," 132.

43. ibid., 124, 131; Jones, Payne, Sukanto, and Partoutomo, "*Trypanosoma evansi.*"

44. Reid, "*Trypanosoma evansi* Control," 219–24.

45. Ware, "India," 37.

46. S. R. Rippon, "Burma," in *History of the Overseas Veterinary Service, Part I*, 70; Eugène Teston and Maurice Percheron, *L'Indochine moderne* (Paris: Librairie de France, 1931), 918; Philippines, *Annual Report of the Bureau of Agriculture 1926* (Manila: Bureau of Printing), 83.

47. A. G. Luckins, "Epidemiology of Non-Tsetse-Transmitted Trypanosomiasis: *Trypanosoma evansi* in Perspective," http://www.icptv.org/Newsletters/Newsletter1.epidemiology.html.

48. Shelagh Lloyd, "Diseases Caused by Protozoa," in M. Horace Hayes, *Veterinary Notes for Horse Owners: An Illustrated Manual of Horse Medicine and Surgery*, ed. Peter D. Rossdale, 17th ed. (London: Stanley Paul, 1987), 433–35; Secretariat of the Pacific Community, "Equine Babesiosis," http://www.spc.int/rahs/Manual/Equine/BABESIOSISE.htm.

49. Tricard, "Le cheval en Indochine," 213; *Encyclopaedie*, 1st ed., vol. 4, 507–8, and 2nd ed., vol. 4, 523.

50. M. R. Montemayor, "Half a Century of Livestock Raising," in *A Half Century of Philippine Agriculture* (Manila: Graphic House, 1952), 277–78.

51. Lise Wilkinson, *Animals and Disease: An Introduction to the History of Comparative Medicine* (Cambridge: Cambridge University Press, 2005).

52. Lise Wilkinson, "Glanders: Medicine and Veterinary Medicine in Common Pursuit of a Contagious Disease," *Medical History* 25, no. 4 (1981): 363–84; Anne Grimshaw, *The Horse: A Bibliography of British Books, 1851–1976* (London: The Library Association, 1982), 33, 112, 116–17, 189; Rhode Island Medical Veterinary Association, "Glanders," 2003, http://www.rivma.org/Glanders.doc.

53. Fletcher, *Tengyueh*, 98; Tricard, "Le cheval en Indochine," 213; Lancaster, "Malaya," 94, 97, 102; Knapen, *Forests of Fortune?* 296; Barwegen, "Gouden hoorns," 132; Mulder, "De economische beteekenis," 325; Bankoff, "Horsing Around," 247.

54. Foreman, *Philippine Islands*, 336, 622.

55. Barwegen, "Gouden hoorns," 133.

56. Ware, "India," 26; Rippon, "Burma," 54, 71; Bankoff, "Horsing Around," 249.

57. Mulder, "De economische beteekenis," 325–26; *Encyclopaedie*, 2nd ed., vol. 4, 523. For numbers imported, see Clarence-Smith, "Horse Trading," 144.

58. Rippon, "Burma," 71.

59. Malcolm C. Roberts, "Diseases Caused by Bacteria," in Hayes, *Veterinary Notes for Horse Owners*, 411, 414; Grimshaw, *Horse*, 112.

60. Cuthbert W. Harrison, *An Illustrated Guide to the Federated Malay States (1923)* (Singapore: Oxford University Press, 1985), 97–98; *Encyclopaedie*, 2nd ed., vol. 4, 524; Bankoff, "Horsing Around," 247–48; Fletcher, *Tengyueh*, 98.

61. Roberts, "Diseases Caused by Bacteria," 413, 679; Charles L. Stoltenow, North Dakota state University Extension Service, "Anthrax," http://www.ext.nodak.edu/extpubs/ansci/livestock/a561w.htm.

62. H. N. Marshall, *Elephant Kingdom* (London: Travel Book Club, 1959) 63; Executive Committee, *Siam in 1930*, 265; Rippon, "Burma," 65–67; Buchy, "L'Institut Pasteur de Nha Trang"; Lancaster, "Malaya," 102; *Encyclopaedie*, 1st ed., vol. 4, 507; Mulder, "De economische beteekenis," 325; Bankoff, "Horsing Around," 247.

63. Rippon, "Burma," 65–67; Philippines, *Annual Report 1924–27*.

64. Roberts, "Diseases Caused by Bacteria," 414–15; Grimshaw, *Horse*, 112.

65. *Encyclopaedie*, 2nd ed., vol. 4, 524; Mulder, "De economische beteekenis," 325.

66. *Institute for Medical Research*, 33; Virginia Thompson, *Thailand: The New Siam* (New York: Paragon Books, 1967), 702; Institut Pasteur de Bangkok,

"Notice descriptive du fonds," http://www.pasteur.fr/infosci/archives/ban1 .html.

67. Ware, "India," 32–37; Indian Veterinary Research Institute Izatnagar, "Muketswar Campus," http://ivri.nic.in/campuses/muketswar.htm.

68. Rippon, "Burma," 50–66.

69. J. Russell Andrus, *Burmese Economic Life* (Stanford, CA: Stanford University Press, 1947), 53–54.

70. *Imperial Gazetteer*, vol. 22, 258.

71. Harrison, *An Illustrated Guide*, 97–98; Clarence-Smith, "Cape to Siberia," 60–61.

72. Lancaster, "Malaya," 90–111; *Institute for Medical Research*, 38–39, 49.

73. Barwegen, "Gouden hoorns," 67–72.

74. Teston and Percheron, *L'Indochine moderne*, 916–19.

75. Buchy, "L'Institut Pasteur de Nha Trang."

76. Mariano Malabanan, "History of the Bureau of Animal Industry," in *Half Century of Philippine Agriculture*, 267–76; Bankoff, "Horsing Around," 248–49, 251, 269.

77. Thompson, *Thailand*, 334.

78. Chumphon Naewchampa, "Socio-economic Changes in the Mun River Basin, 1900–1970," in *The Dry Areas of Southeast Asia: Harsh or Benign Environment?* ed. Hayao Fukui (Kyoto: Center for Southeast Asian Studies, 1999), 223.

79. Executive Committee, *Siam in 1930*, 264–65.

80. Andrus, *Burmese Economic Life*, 54–55; Malabanan, "History of the Bureau," 271.

81. Lucien Tichit, *L'agriculture au Cambodge* ([Paris]: Agence de Coopération Culturelle et Technique, 1981), 297, 320.

82. Pieter Hoekstra, *Pardenteelt op het Eiland Soemba* (Batavia: Drukkerij V/H John Kappee, 1948), 131.

83. Lancaster, "Malaya," 115; *Institute for Medical Research*, 309–10.

84. Luckins, "*Trypanosoma evansi* in Asia," 138.

85. Executive Committee, *Siam in 1930*, 264–65.

86. Lewis E. Gleeck Jr., *Laguna in American Times: Coconuts and Revolucionarios* (Manila: Historical Conservation Society, 1981), 44, 51–62.

87. Dean C. Worcester, *The Philippines Past and Present* (New York: Macmillan, 1930), 420.

88. Rippon, "Burma," 50–52, 54–56, 62–65.

89. Ware, "India," 20–23, 26.

90. Lancaster, "Malaya," 105, 109, 117.

91. *Institute for Medical Research*, 44.

92. Teston and Percheron, *L'Indochine moderne*, 916.

93. Barwegen, "Gouden hoorns," 70–72.

94. Rippon, "Burma," 71–72; *Institute for Medical Research,* 73–74.

95. Barwegen, "Gouden hoorns," 135–37.

96. J. L. Kipling, *Beast and Man in India: A Popular Sketch of Indian Animals in Their Relation with People* (London: Macmillan, 1921), 194–95.

97. Lancaster, "Malaya," 103; *Institute for Medical Research,* 17–19, 23–26.

98. Bankoff, "Horsing Around," 248.

CHAPTER 8

"They Give Me Fever"

East Coast Fever and Other Environmental Impacts of the Maasai Moves

Lotte Hughes

IT IS widely acknowledged that the Maasai of British East Africa (BEA, later renamed Kenya) were relieved of the best part of their territory in the 1900s and moved at gunpoint into reserves, in order to free up the highlands for white settlement.[1] European financial interests, suggest G. H. Mungeam, M. P. K. Sorrenson, Diana Wylie, and others, were the main driver behind moves in the period 1904–5 and from 1911 to 1913, involving upward of twenty thousand people and at least 2.5 million livestock.[2] But by the time of the second move, from Laikipia in the highlands to what is now western Narok District, was there a related but less obvious motive, linked to settler pressure on government to take action on East Coast fever (ECF) and grant white farmers more land in ECF-free areas? My research indicates that ECF was a key factor in the second move, which is barely mentioned in the written literature but is foregrounded in Maasai oral testimony. It was my elderly informants' insistence on linking ECF to the moves and land losses, within minutes of beginning to talk about them, that alerted me to this possibility. Oral claims, albeit problematic, can in part be verified in veterinary and other archival sources.

The existing historical literature on the moves does not cover the disease angle or the environmental impacts. Neither does it include Maasai oral testimony on this or any other aspect of the moves and related events, which is revelatory. Mungeam, Sorrenson, and T. H. R. Cashmore focus largely on official policy toward the Maasai, settler influence upon it, the differences and strained relations between the protectorate government and the Colonial Office (CO), and what happened as a result. R. L. Tignor covers these issues in less detail, mentioning in passing the incidence of ECF in the Southern Maasai Reserve.[3] Wylie's interest is in the renegade British civil servants who challenged this policy and the wider human rights networks to which they belonged. Richard Waller has produced rich histories of the Maasai but has not researched the moves themselves.[4] Colonial civil servant George Sandford, author of an official history of the Maasai reserves, described the diseases that afflicted Maasai livestock before and after the moves, but his account cannot be considered scholarly or independent, even though it is a valuable source. Polemical books published in the 1920s by Dr. Norman Leys and his friend William McGregor Ross (Wylie's subjects) are also important contemporary sources, not scholarly histories.[5] Therefore, my analysis fills a gap, while also complementing— and casting new light upon—existing material.

Initially called African Coast fever, ECF was first diagnosed in the protectorate in 1904, in a herd of cattle brought from the Kilimanjaro area of German East Africa to Nairobi.[6] A disease of cattle caused by the protozoan parasite *Theileria parva*, carried by the brown ear tick, *Rhipicephalus appendiculatus*, it hampered early European settlers' attempts to establish dairy and beef ranches.[7] Lord Delamere, who settled in the protectorate in 1903, lost nearly all his young stock to it at Njoro in the Rift Valley. Settlers were panicked by news of the devastation being caused by ECF in Rhodesia and the Transvaal.[8] Winston Churchill was briefed about ECF when he visited BEA in 1907 as a Colonial Office minister and suggested remedies—wire fencing and quarantine—in his account of that journey.[9] By 1909–10, ECF was seriously worrying settlers, who lobbied the new governor, Percy Girouard. He wired the CO on behalf of farmers who were demanding more land in so-called clean areas.[10] Later, the CO discovered that Girouard had promised land on Laikipia to settlers before the Maasai had "agreed" to vacate it and lied to the CO about these pledges—a deception that led to his downfall.

According to official correspondence between London and BEA, the highlands were free of ECF at this stage. In February 1910, Girouard begged the CO for more money to prevent ECF's spreading to the highlands. A year later, Laikipia was apparently still free of the disease when Acting Chief Veterinary Officer Francis Brandt visited the Northern Reserve to investigate an outbreak of bovine pleuropneumonia (BPP) in Maasai herds and to find out whether ECF was known there. Maasai told him they knew it existed in the south of the country, but they had never experienced it on Laikipia.[11] Plans to move the Maasai for a second time—which had been in the air since at least 1908—were immediately accelerated, which was surely no coincidence. The move would have happened anyway, for a variety of reasons, but I suggest that ECF was central. Also, having placed African reserves in quarantine from 1908, the government had a perfect excuse for preventing Maasai leaders from selling cattle in order to raise the necessary legal fees to bring a lawsuit. This was one of many obstructions designed to derail Maasai attempts to regain Laikipia in the courts. The so-called Maasai Case went ahead in 1913 but was lost on a technicality.

Maasai elders in western Narok still talk with passion about the effects of the second move on the health of humans and herds. They describe the impact of the move in pathological terms, believing that the British deliberately sent them "to that land where *ol-tikana* is" in order that they might die there. (The word means both ECF and human malaria.)[12] They claim that they and their herds succumbed to diseases in the Southern Reserve that were either unknown or not prevalent in their former northern territory, particularly Laikipia, and that they have been blighted by sickness ever since. They insist that the land they were moved to was not only grossly inferior to Laikipia with regard to water supplies, pastures, and disease vectors but that the new environment infected and killed them. It was literally deadly. Some go further and insist there was no disease in Laikipia. In the collective oral mythology, Laikipia is seen as paradise, its sweetness constantly compared to the bitterness of the south.

The Maasai's quantitative land losses to the British in this period are well known. But their qualitative losses, in terms of the richness of their northern habitat and their alleged propensity to disease in their new environment, have not previously been examined in detail. Isaac Sindiga has described how "colonial intervention in Maasailand led to the breakdown of traditional ecosystems" and attributes the subsequent severe degradation and pressures on land in Kajiado and Narok Districts to a process begun in 1904.[13] Other scholars have made similar remarks, linking the early land losses to environmental degradation, increased vulnerability to drought, and other long-term challenges.[14] This chapter attempts to explore the qualitative "before" and "after" of the land alienation in more depth.[15]

In theory, the area of land to which the Maasai were moved, the 4.4 million-acre western extension added to the Southern Reserve in 1911, seems generous until one examines its quality. (The reserve as a whole was nearly ten million acres; one official admitted that "the majority of this is waterless."[16]) But it is not simply a matter of the quality of land. Such is the Maasai dependency on livestock and their total identification with cattle—*en-kishu* means both cattle and the Maasai as a people—that cattle disease is inextricably linked to human health and is spoken of almost interchangeably with that of humans. Furthermore, the grievances of these migrants must be seen in the context of acclimatization over time and space. Other Maasai were already living in this area when the "northerners" arrived, so it cannot be dismissed as an environment in which Maasai could not survive.[17] The point is that the newcomers were unfamiliar with it; were nonresistant to certain infections; and in the interregnum between arrival and acclimatization, when some resistance developed, both humans and stock suffered acutely. It is this suffering that colors people's memories of the move and what happened immediately afterward.

One could attempt to establish whether there is any scientific or biomedical basis for Maasai claims that the Northern Reserve was effectively ECF-free and that a deliberate "move-to-kill" policy was driven by administrators' knowledge of the presence or absence of disease, particularly ECF, in the two environments. There is some compelling evidence to support the first of these claims, and some colonial officials certainly "subscribed to a Malthusian view of disease and drought as the natural regulators of the [African] stock population."[18] But the search for scientific evidence is also an unsatisfactory exercise, in part because early scientific data simply do not exist and because it involves comparing like with unlike: to put it crudely, a Western scientific view of disease that is rooted in diagnostics and laboratory experiment versus a more holistic indigenous view that regards "dis-ease" as a natural part of life. Most importantly, the subject is larger than scientific: it concerns disease as a metaphor for colonial encounters and what these produced in social and other terms. In this case, I argue that ECF has come to represent—for the older generation of Purko Maasai, at least—infection by colonialism, and it is their conceptualization that interests me. Therefore, I aim to examine what scientific evidence there was in tandem with *perceptions* of disease and socioenvironmental health, confining my focus largely to the Purko section who bore the brunt of the second move and to ECF.

The Classifications "Clean" and "Dirty"

In the 1900s veterinary officials divided BEA into "clean" and "dirty" areas, according to the incidence of ECF. In 1911–12, Maasai cattle on Laikipia were said to be clean. The Rift Valley, home to most of the imported settler stock, was also ECF-free at that time, while the Southern Reserve was classified as dirty but only "slightly infected." In the official lexicon, "clean" and "dirty" generally referred to areas of healthy or sick cattle populations and areas of white or African settlement, respectively, and underpinned administrative action to keep them separate. This ignored the fact that some settler farms in supposedly clean areas were actually "dirty"; outbreaks of ECF in the "clean" Rift "were not uncommon."[19] The administration tended to view "native" cattle as inherently inferior and diseased—although veterinarians recognized their resistance to some diseases, including ECF. It sought to keep "native" stock separate from imported, pure-bred settler stock through compulsory measures such as fencing, quarantine, and the removal of squatters from European farms.

For their part, the "northern" Maasai also regarded the Southern Reserve as "dirty." They believed, both before and after the moves, that they were deliberately moved to inferior and waterless pasture infested with ticks and tsetse fly, where both human and stock diseases were rife. The plaint in the 1913 Maasai Case specifically stated, "The Southern Masai Reserve to which the stock of the Masai is being moved is infected with East Coast fever."[20]

Comparing the Two Habitats

The Maasai use the word Entorror to refer to the whole of their former northern grazing grounds, but most recently to Laikipia. In leaving Entorror, the Maasai swapped a territory that was generally higher, cooler, and wetter for largely semi-arid plains with no appreciable swamps and fewer highland drought refuges. Sindiga notes,

> Only a small fringe of territory had good grass and water all year round. Nearly all permanent streams were controlled by European settlers. The rest of the territory . . . was either without water, or contaminated by disease, or in European control. At their level of technology the Maasai could not readily use a total of 51 per cent of their reserve.[21]

Compared to the north, there were few accessible forests, apart from those on the Mau escarpment, southwest of it, and at Chepalungu in Trans-

Mara. But the first and last of these were truncated by the reserve boundary and were far from the best plains grazing, and the best part of Chepalungu was out of bounds. Although Laikipia was not ideal and not large enough, it offered a wider variety of options than the western extension of the Southern Reserve, particularly at a time when policing was relatively relaxed and herders were allowed to break boundaries during droughts. Leys compared the two reserves in this way: "No European in the country imagined for a moment that the Masai on Laikipia wished to leave it. The area, though small, is as fine a piece of country as there is in Kenya, with rich soil and perennial streams, vastly superior in every way to the country south of the Rift Valley."[22]

In leaving the highlands, the "northern" Maasai lost the wide choice of habitat they had once enjoyed; before 1904, of course, their range was even wider. Transhumant pastoralists make use of ecological niches. If they are free to do so and control the territory, they move in and out of these niches according to seasonal need and constantly stress the balance to be achieved in rangeland and stock management between highland, dry-season grazing (*osupuko,* the drought refuges) and lowland, wet-season grazing (*ol-purkel*). Moreover, each section has its own osupuko and ol-purkel; they cannot easily find alternatives on moving to a region already occupied by others.

Though early scientific studies do not exist, one can get a fairly good idea of the natural resources of precolonial Maasailand from travel texts, notably the writings of Scottish geologist Joseph Thomson and naturalist/ administrator Harry Johnston and later from official reports written before mass European settlement began. (For reasons of space, I shall not discuss Thomson and Johnston's work here).[23] To paraphrase reports written in the 1900s by administrators, foresters, and agricultural officers, the Rift, Mau, and Laikipia were said to have the richest pastures in the country, drought-resistant species of grass, good soils, valuable forests, and abundant water supplies. The Rift grazing was said to be superb precisely because Maasai cattle had grazed it for years. Charles Hobley, then assistant deputy commissioner, toured Laikipia in June 1904 to check its suitability as a reserve for Maasai who were soon to be evicted from the Rift. Fine water supplies, rich belts of forest, green rolling plains, "magnificent grazing country"—Laikipia had it all.[24] Also, he saw little big game outside the forests. His report confirmed what the Maasai claim today—that there was plenty of water, wonderful pasture, fewer large game animals than on the southern plains, and, therefore, fewer tick hosts.

In the south, by comparison, the immigrants faced stiff competition from wild animals for pasture and water. They also competed for resources with Maasai sections already living there, principally the Loitai and Siria, together with Purko who had moved south from the Rift Valley in 1904, instead of relocating to Laikipia.[25] There were larger numbers of predators and disease carriers. For example, wildebeest migrating north from the Serengeti transmitted malignant catarrhal fever to cattle; even by the 1960s, the death rate was more than 95 percent and no treatment or vaccine was available.[26] The only permanent rivers were the Mara, its tributary the Talek, and the southern Uaso Nyiro; other watercourses dried up in drought years. Fewer highland drought refuges were available to the incomers. Much of the grassland was superb, but tsetse and ticks rendered enormous areas useless. Robertshaw and Lamprey describe how "the Maasai were confined, by tsetse infestation of the Mara Plains, to [the Lemek] area for the greater part of [the twentieth] century."[27] In the 1930s, research on ticks and tsetse by government entomologist Aneurin Lewis provided crucial evidence that millions of acres of otherwise good grazing were rendered out of bounds to the Maasai.[28]

The Maasai Version of Events

When asked about the forced moves, Purko elders insist that the British told the Maasai, "*Shomo Ngatet mikiwa ol-tikana.*" This means, "Go to the south and may malaria/ECF kill you there." Interviewees made these claims before any question was asked about the incidence of disease in the two reserves. I spoke to several members of the Il-Terito age-set who were born in Laikipia and took part in the moves as small children. The majority of my sixty-four interviewees made similar claims. For example, one of my oldest informants, the late Muiya Ole Nchoe of Lemek, said, "Ole Gilisho was told by the white man: 'Get out of Entorror because even us, we want to put our cattle and people here where there are no diseases.' The land was very suitable for the en-kishu. [There were] no diseases for both people and cattle, no diseases completely."[29]

Other respondents refuted this, saying there had been *ol-odua* (rinderpest), *ol-kipiei* (BPP), and *empuruo* (anthrax) on Laikipia—but nothing compared to how ECF swept through Maasai herds once they moved south. They also alleged that serious human sickness was virtually unknown in Entorror, and the migrants were highly susceptible to infections in their new environment.[30] There are many examples in the testimonies I collected of a belief in a deliberate British action to exterminate the Maasai. A collective folk memory has evolved in which this idea is central.

Official Views and Interventions

I have suggested that British administrators' knowledge of the effective absence of ECF on Laikipia was a factor behind the second move. By "effective," I mean that while the disease may have been present, genetic or acquired stock resistance (immunity may be too strong a term) rendered it relatively harmless. The available evidence indicates that Maasai herds on the plateau were resistant to ECF or to a particular strain of it. The ability of indigenous Zebu and Zebu crosses to acquire immunity to ECF was known by 1910. Resistance was acquired through an attack in early calfhood; some adult cattle then became carriers. Problems would only have arisen when the cattle moved south and met at least four new conditions: exposure to infected country en route, exposure to other strains, exposure to larger numbers of game (particularly buffalo), and higher concentrations of cattle in a more restricted area of grazing.

On the first and third points, driving stock through forest corridors was very risky because of the numbers of ticks and buffalo found there. It was common knowledge, said informant Desmond Bristow (former farm manager to cattle baron Gilbert Colvile and the "second" Lord Delamere, fourth Baron Cholmondeley), that the stock routes leading south from Thomson's Falls were heavily tick infested.[31] Ticks thrive in long, coarse grass, where humidity facilitates egg laying and the molting of larva and nymph, and they multiply after the rains. Forest corridors during and after the rains were the ideal environment for tick populations, lying in thick, shaded vegetation protected by overhanging trees. Also, a variant of ECF is bovine cerebral theileriosis or corridor disease, so-called because it is picked up in forest corridors. It is caused by a very similar protozoan parasite, *T parva lawrencei*, and transmitted to cows from buffalo by the same vector tick as ECF. Cattle resistant to or immunized against *T parva parva* often cannot withstand *T parva lawrencei*. The strong possibility that cattle were exposed while moving to both parasites, bearing disease strains to which they lacked resistance, must be factored in.

As for the risks arising from higher concentrations of cattle in a more restricted area, Francis Brandt wrote at the time of the move:

> The Masai . . . with unlimited grazing, are accustomed immediately on the appearance of any disease to move their cattle to fresh grazing grounds, with, in the case of East Coast fever, a loss of only one or two head of cattle. . . . Infection in the shape of infected ticks is left behind ready to attack the next

herd of cattle which pass. In this way, so long as the country is under-stocked, the losses are inappreciable, but in the event of an excess of cattle being grazed over a limited area an epidemic of East Coast fever would probably occur.[32]

Here was the recipe for disaster, and the veterinarians were foretelling it before the Maasai moved. However, they did not see the Southern Reserve as limited in size. Another possibility, which may have disguised the presence of ECF on Laikipia, is enzootic stability. According to this theory, ECF existed, but there was little clinical manifestation of it. Maasai cattle acquired resistance, and disease would only occur after susceptible animals moved into the enzootic areas or after enzootic areas were extended into contiguous but previously tick-free nonenzootic areas. The extension areas developed more grass and other ground cover after the rains, new populations of the vector tick built up in the new vegetation, and epizootics broke out in susceptible cattle. By this reckoning, the best thing to do with a tick-borne disease was nothing—except allow animals to develop immunity. The Maasai knew this then. The veterinary authorities also recognized the role of endemicity in the development of immunity.[33] Some veterinarians now say this is the ideal strategy.[34] Though vaccines and drug treatments are now available, killing ticks by dipping cattle is still the main preventative measure.

Veterinary interest in Maasai herds on Laikipia had been cursory before 1911 and increased only when the second move was mooted. Officials were concerned about the threat posed to settler stock since the Maasai were to be driven south across their farms. The main diseases among Maasai cattle between 1904 and 1912 were BPP and gastroenteritis (more likely to have been rinderpest). Rinderpest, BPP, black quarter, *engamuni,* and *m'benik* (now known as ephemeral fever) were noted in 1911–12 and redwater in 1913. Rinderpest was thought to have existed on Laikipia "for many years in an endemic form," striking down mostly young animals and leaving adult survivors immune. Gastroenteritis "swept through" the herds in 1909 and 1910, killing as many as one thousand head of cattle in some bomas (stockades); an estimated fifteen thousand cattle and calves died before the disease "disappeared." The Maasai insisted that it was actually rinderpest, and afterward they considered their herds immune to it; it seems they were right on both counts. The presence of fatal gastroenteritis (rinderpest) in the herds at the end of April 1910 "suddenly" forced the authorities to delay the start of the move. But two other events also caused this delay: the Colonial Office had cabled Girouard with orders to halt the move until the

Maasai agreed to do so, and age-set leader Parsaloi Ole Gilisho had changed his mind about moving south.

Long after the Maasai had gone, Laikipia remained an officially ECF-free area. So far as I can tell, the first reported outbreak in Rumuruti was in 1919–20, which was said to have originated in an animal from West Kenya.[35] The evidence points to its likely introduction by settlers, who spread it by rail and by road-transport oxen. Today it is regarded by farmers as a constant scourge.

Conditions in the Southern Reserve

"Many cases of East Coast fever have come to notice from this reserve," wrote Chief Veterinary Officer Robert Stordy in 1910, though he went on to say that large areas were "sparsely infected" because of the Maasai habit of moving their animals away from infected grazing grounds, which had over time become "automatically clean."[36] Veterinary Officer Bill Kennedy reported that rinderpest was prevalent in the Southern Reserve before the move began, but it had hit the incoming herds particularly hard because many younger cattle were susceptible. A "serious epizootic" broke out, and by December 1913 rinderpest was "very widespread."[37]

The following year a larger crisis loomed. World War I was fought in East Africa on the Maasai southern front of the border with German East Africa. The war effort and preparations for it involved major livestock movement—not only an increase in ox-drawn transport but also in the mass movement of slaughter cattle acquired from the Maasai to feed the troops. This helped to spread ECF in Maasailand and elsewhere. Quarantine and restrictions on livestock movement were both eased during this period, and many white farmers even took their oxen into battle. By 1917, all these factors had led to outbreaks in Nakuru, the Limuru "clean area," Naivasha, and the Southern Reserve. By the time war ended, officials were seriously concerned about the rising incidence of stock disease in the reserve. The officer in charge reported "upwards of half a million deaths" in 1919–20 from tick-borne diseases, BPP, and rinderpest.[38] In January 1918, an outbreak of rinderpest in Narok District killed nearly all the calves in certain villages. In 1922, it killed between 60 and 100 percent of all cattle in the Narok and Loitokitok areas. Concerns about the spread of BPP led to the first veterinary laboratory being built in the reserve in 1918. Kennedy, now acting chief veterinary officer, traveled to Narok to see what could be done about BPP. To inoculate the three-quarters of a million or so Maasai cattle in the reserve would require a staff of at least eight veterinarians and fifty stock inspectors. He only had twelve veterinarians and eleven stock inspectors in the whole protectorate.[39] Consequently, no action was taken.

Sandford gave a full and frank account of the diseases to which Maasai stock succumbed up to 1919. ECF and BPP were the most serious diseases in the reserve, which was placed in continuous quarantine from 1916 following major BPP outbreaks. Quarantine restrictions on stock exports were not lifted until 1935. But ECF caused "by far the highest mortality" of any disease. In August 1914 it had been confined to an area near Ngong, the Sotik border, and Trans-Mara. Since then it had spread rapidly. Most significantly, Sandford noted, "Masai cattle appeared to have bred a certain degree of immunity to the disease, but the Officer-in-Charge was inclined to think that the cattle which had come from Laikipia were less immune than those which had previously resided in the Southern Reserve."[40] He blamed the Maasai for not preventing it through dipping and put this down to superstition and laziness. But my informants said that the Maasai initially resisted dipping because of their distrust of Europeans and also because they hated the trauma it caused their beloved cattle, which were "beaten severely to force them to jump."[41]

Were the authorities trying hard enough to treat and prevent Maasai stock disease in this early period? Leys thought not and scathingly summed up the disparity between veterinary attention to white and black pastoralists: "The Veterinary Department professes to work for the benefit of European and African stock-owners alike. The claim is sheer nonsense. Nine-tenths of the Department's work consists of free preventative and curative treatment given to the property of Europeans, who own, according to official returns, only 5 percent of the stock in the country."[42] He noted a glaring anomaly: veterinary concerns about disease in "native" reserves, and their classification as "dirty," did not match the numbers of vets assigned to tackling it.[43] The fact that vets were employed almost exclusively in the European areas was freely admitted in annual reports between 1911 and 1924, and the department made plain it had no intention of tackling disease in African reserves. For instance, the 1911 report stated: "Eradication in the vast native reserves where East Coast fever is endemic is not to be thought of, even were it possible." It simply used continuous quarantine of the reserves to stop disease spreading.

Lewis's Study of Ticks

It was not until 1934, with the publication of Aneurin Lewis's pioneering study of ticks in the Southern Reserve, that a clear picture emerged of the extent of both tick infestation and other challenges to stock in Maasai country.

This is a very rich piece of work, based on research in 1932–33, when Lewis was attached to the Veterinary Research Laboratory at Kabete near Nairobi. His findings confirm Maasai claims about the prevalence of ECF in the reserve and how confinement to reserves prevented people from employing age-old coping mechanisms—which involved, in part, moving away from tick-infested pasture.

Lewis's description of the incidence and effects of ECF among Purko herds in western Narok leaves no doubt about the horrors of ol-tikana there:

> The writer witnessed an outbreak of disease among adult cattle which swept away whole herds of hundreds of cattle. In one boma at Aitong, 300 adult beasts died within fourteen days of their return from the Mara.... Dead bodies of sheep were strewn along the routes from Mara bridge to Engoregori. This is also true of the cattle. Indeed the vultures, the hyaenas, jackals and other scavengers could not cope with the abandoned carcases [*sic*]. Examinations of the dead animals and of numerous blood and gland smears proved the cause of death to be East Coast fever.[44]

Lewis went on to claim: "Certain large areas of the Masai country are unsuitable for all stock; others are useful only for sheep, while still others are totally uninhabited by man or domestic beast." The reasons for this were multiple: few permanent water sources; seasonal fluctuations in water supply, which meant certain areas were only useable for part of the year; lack of grazing; the presence of tsetse fly and ticks; and fear of disease in areas where rinderpest and smallpox had previously decimated communities. Where grazing was good, it was often rendered useless. He also saw a link between weakened cattle and their greater susceptibility to ECF, noting "a tendency for ill-conditioned, unhealthy and sick animals to become more liable to the attacks of ticks. Whether it is due to the conditions of the beast and lack of resistance, or to the fact that such animals, by frequently resting often provide more time and opportunity for attack by ticks, it is difficult to say."[45] One may speculate that Maasai cattle weakened by the long march south from Laikipia arrived in a more susceptible state. The move also entailed frequent stops and starts, which would have increased the likelihood of attack by infected ticks en route. Modern veterinary opinion supports this idea.

Lewis plotted on a map the main dry season migration routes of cattle in the reserve. Another map showed shaded areas where the chief agent of ECF, *R. appendiculatus*, thrived. Lewis noted that Maasai will venture into areas they normally avoid when faced with a serious lack of water and grazing, with one obvious result:

> These movements are towards and into areas infected with ECF. . . . When adverse conditions are at an end . . . the Masai with the remainder of their stock wander back to their homes, away from the ECF infected areas. Obviously the stock infested in these areas with *R. appendiculatus* carry this species of tick—and others—to uninfected areas. Thus, bit by bit, new areas of the reserve become infested with this tick.[46]

DESPITE SOME anomalies, there is a solid basis for Maasai belief in the intrinsic healthiness of their former grazing grounds and the comparative health of their herds on Laikipia compared to what befell them in the Southern Reserve. There is compelling evidence that ECF was either unknown on Laikipia before 1911 or, more likely, that Maasai herds were resistant to the strains that existed there, and enzootic stability could be maintained. Cattle succumbed to other strains on moving to unfamiliar territory. Also, nomadic coping strategies involving movement away from tick-infested pastures kept ECF at bay or under control. The stress of the long march south would have weakened the livestock and increased its vulnerability to disease. On arrival, stock losses were exacerbated by poor veterinary support, at least in the early years. What Sandford and other officials dismissed as baseless griping reflected very real and life-threatening concerns. European settlers coveted Laikipia precisely because it appeared to be free of ECF and malaria, and they lobbied the government in order to get it. Official knowledge of the effective absence of ECF on Laikipia at this time was, according to this evidence, a crucial factor in government plans to oust Maasai stock keepers and replace them with Europeans. Other scholars are right to identify European financial interests as a major force behind the moves, as healthy livestock meant healthy bank balances.

However, some elders' claims that no stock diseases existed on Laikipia when the Maasai lived there are patently untrue. Official records describe the high incidence of disease in the 1900s, which delayed the start of the second move. It is probable that Maasai informants were referring to diseases new to them since they moved south, such as "new" strains of ECF. Also, the collective folk memory of late nineteenth-century epidemics (rinderpest, BPP, and human smallpox) is so appalling that it has eclipsed that of smaller outbreaks of disease in the period up to 1912.[47] By comparison, any lesser calamities were relatively easy to dismiss. More broadly, for a variety of sociopolitical reasons Maasai recollections of Laikipia (particularly by members of the Purko and Laikipiak sections) are rose-tinted—it has acquired the status of a lost Eden, free of disease and white ranchers, the like of which will never be seen again.[48]

Notes

1. These events and their repercussions are described in my doctoral dissertation "Moving the Maasai: A Colonial Misadventure" (University of Oxford, 2002), a revised form of which was published as a book of the same title (Basingstoke, England: Palgrave Macmillan, 2006). The wider environmental impacts of the moves are described in chapter 5 of the latter. Maasai is the correct spelling, but Masai is used when citing colonial records. British East Africa was also called the East Africa Protectorate.

2. Estimates from G. R. Sandford, *An Administrative and Political History of the Masai Reserve* (London: Waterlow & Sons, 1919); he does not give a definitive total. Other sources on European settlement, the moves, and opposition to them include T. H. R. Cashmore, "Studies in District Administration in the East African [*sic*] Protectorate, 1895–1918" (PhD diss., University of Cambridge, 1965); Norman Leys, *Kenya* (London: Hogarth Press, 1924); G. H. Mungeam, *British Rule in Kenya, 1885–1912* (Oxford: Clarendon Press, 1966); W. McGregor Ross, *Kenya from Within* (London: George Allen & Unwin, 1927); M. P. K. Sorrenson, *Origins of European Settlement in Kenya* (Nairobi: Oxford University Press, 1968); R. L. Tignor, *The Colonial Transformation of Kenya: The Kamba, Kikuyu and Maasai from 1900 to 1939* (Princeton, NJ: Princeton University Press, 1976); Diana Wylie, "Norman Leys and McGregor Ross: A Case Study in the Conscience of African Empire, 1900–39," *Journal of Imperial and Commonwealth History* 5, no. 3 (1997): 294–309.

3. Tignor, *Colonial Transformation,* 36, citing a report on Trans-Mara by Rupert Hemsted, officer in charge of the Southern Reserve, 5 May 1913. Hemsted wrote, "I suspect the tick which carries East Coast fever is fairly prevalent" (orig. at 65). Copy in the Lewis Harcourt Papers, Bodleian Library, Oxford.

4. R. Waller, "The Maasai and the British, 1895–1905: The Origins of an Alliance," *Journal of African History* 17, no. 4 (1976): 529–53, provides essential background to the Maasai moves; idem, "Tsetse Fly in Western Narok, Kenya," *Journal of African History* 31, no. 1 (1990): 81–101, investigates the spread of tsetse, a vector of trypanosomiasis, in the area to which the "northern" Maasai moved.

5. Leys was a medical doctor who fully grasped the implications of forced migration for both human and animal health. But he was ejected from BEA in 1912–13, before he could chronicle the effects of the moves on health.

6. R. A. I. Norval, B. D. Perry, and A. S. Young, *The Epidemiology of Theilerosis in Africa* (London: Academic Press, 1992), 48.

7. *T parva* sporozoites are injected into cattle by infected vector ticks during feeding. Of three subtypes, the first two are highly pathogenic and can cause high levels of mortality: *T parva parva,* largely transmitted between cattle; *T parva lawrencei,* mainly transmitted from buffalo to cattle; and *T parva bovis,* transmitted between cattle. Symptoms include fever, swelling of the lymph nodes, loss of condition, labored breathing, and nasal discharge. *The Merck Veterinary Manual,* http://www.merckvetmanual.com.

8. P. F. Cranefield, *Science and Empire: East Coast Fever in Rhodesia and the Transvaal* (Cambridge: Cambridge University Press, 1991), 225, 181–222.

9. Winston Churchill, *My African Journey* (London: Hodder and Stoughton, 1908), 40.

10. Girouard to CO, 1 and 4 April 1910, Desp. 175, CO 533/71, NA.

11. *Laikipia District Annual Report for the Year Ended 31 March 1911,* LAK/1, Kenya National Archives (KNA).

12. Frans Mol, *Maasai Language and Culture Dictionary* (Limuru, Kenya: Kolbe Press, 1996), 387.

13. Isaac Sindiga, "Land and Population Problems in Kajiado and Narok, Kenya," *African Studies Review* 27, no. 1 (1984): 27.

14. See, for example, Tignor, *Colonial Transformation,* 38; Marcel Rutten, *Selling Wealth to Buy Poverty: The Process of the Individualization of Landownership among the Maasai Pastoralists of Kajiado District, Kenya, 1890–1990* (Saabrücken: Verlag Breitenbach, 1992), 8.

15. For more detail, see Hughes, *Moving the Maasai,* chap. 5.

16. H. R. McClure, "District Records for the Guidance of the Officer Administering the Masai Southern Reserve," undated manuscript, British Institute in Eastern Africa, Nairobi.

17. The Maa-speaking groups in Kenya are divided into autonomous socio-territorial sections called *il-oshon* (sing. *ol-osho*). Today there are some twenty-two sections (though some scholars give a lower number), of which the Purko section is one of the largest.

18. Richard Waller, "'Clean' and 'Dirty': Cattle Disease and Control Policy in Colonial Kenya, 1900–40," *Journal of African History* 45, no. 1 (2004): 46. I wrote the section of my dissertation upon which this chapter is based at least three years before this article was published; hence, my ideas are not derived from it.

19. Norval, Perry, and Young, *Epidemiology of Theileriosis,* 51.

20. Pleadings, Civil Case No. 91 of 1912, in Conf. 11, 16 Jan. 1913, CO 533/116, NA.

21. Sindiga, "Land and Population Problems," 27. Leys first made the point about streams rising on European-controlled land on the Mau. See Leys, *Kenya,* 2nd ed. (London: Hogarth Press, 1925), 107, 110.

22. Leys, *Kenya* (2nd ed.), 104.

23. Joseph Thomson, *Through Masai Land* (London: Sampson Low, Marston, Searle and Rivington, 1885). Harry Johnston's publications include *The Uganda Protectorate* (London: Hutchinson, 1902).

24. C. W. Hobley, "Journey from Naivasha to Baringo and the Laikipia Highlands," 24 June 1904, Enc. 1 in Desp. 493, 22 July 1904, FO 2/838, NA.

25. Waller, "Tsetse Fly in Western Narok," 95.

26. Walter Plowright, "Inter-relationships between Virus Infections of Game and Domestic Animals," *East African Agricultural and Forestry Journal* 33 (June 1968): 262.

27. Peter Robertshaw with Richard Lamprey, "The Research Area," in *Early Pastoralists of South-western Kenya,* ed. Peter Robertshaw (Nairobi: British Institute in Eastern Africa, 1990), 11. See also Waller, "Tsetse Fly in Western Narok."

28. E. A. Lewis, "A Study of the Ticks in Kenya Colony, Part III: Investigations into the Tick Problem in the Masai Reserve," *Bulletin No. 7 of 1934* (Nairobi: Government Printer, 1934); idem, "Tsetse-Flies in the Masai Reserve, Kenya Colony," *Bulletin of Entomological Research* 25, no. 1 (1934): 439–55.

29. Right-hand or senior circumcision group of Il-Terito age-set, interviewed 2000. Ole Gilisho was an important Purko age-set spokesman (leader) who tried to resist the second move and initiated the 1913 lawsuit. All Maasai males are members of age-sets, which are like rungs on the ladder of life, into which boys are initiated at circumcision.

30. See Hughes, *Moving the Maasai;* references to human health have been cut from this chapter.

31. Desmond Bristow, interview by author, Solio Ranch, Laikipia, Kenya, 2000.

32. *Annual Report of the Veterinary Department for the Year 1911–1912* (Nairobi: Government Printer, 1913), 20.

33. For example, see *Annual Report of the Veterinary Department for the Year 1912–1913* (Nairobi: Government Printer, 1913).

34. My thanks to Dr. Glyn Davies for explaining these points and many others.

35. *Annual Report of the Acting Chief Veterinary Officer for the Year Ending 31st March 1920,* appendix B in *Department of Agriculture Annual Report 1919–1920* (Nairobi: Government Printer, 1921).

36. *Annual Report of the Veterinary Department for the Year 1909–1910* (Nairobi: Government Printer, 1911). Quoted in Lewis, "Study of the Ticks," 9.

37. Report of William Kennedy, 21 April 1913 in *Annual Report of the Veterinary Department for the Year 1912–1913*, 51.

38. *Annual Report of the Acting Chief Veterinary Officer Year Ended 31 March 1920*, 24, quoting the officer in charge of the Southern Reserve.

39. *Annual Report of the Veterinary Department for the Year 1917–1918* (Nairobi: Government Printer, 1919).

40. Sandford, *Administrative History*, 64–65.

41. Information supplied by Dickson Kaelo from interviews with elders Ndeyo Ole Yiaile and Konana Ole Kereto.

42. Leys, *Kenya* (2nd ed.), 105n. He discusses ECF briefly, 122.

43. Leys, *A Last Chance in Kenya* (London: Hogarth Press, 1931), 95.

44. Lewis, "Study of the Ticks," 48–49. While sheep do not get ECF, Nairobi sheep disease is transmitted by the same tick, and Lewis implies that it caused these deaths.

45. Ibid., 24, 60.

46. Ibid., 28.

47. For the epidemics and their aftermath, see Waller, "Maasai and the British."

48. This is discussed more fully in the conclusion to my book and in my article "Malice in Maasailand: The Historical Roots of Current Political Struggles," *African Affairs* 104, no. 415 (2005): 207–24.

CHAPTER 9

Animal Disease and Veterinary Administration in Trinidad and Tobago, 1879–1962

Rita Pemberton

FROM THE advent of European control of the Caribbean, there was a preoccupation with plants, the cultivation of which dominated the agricultural sector. However, animals have also played important roles in the development of these territories. They featured in the food and rituals of the indigenous peoples and were central to European colonial activity. The cultivation of the major crops was dependent upon imported animals. Since there is no indigenous species raised as livestock in the region, all reared species have been derived from imported breeds.[1] The scattered references to animals in the existing literature focus on these animal imports. Alfred Crosby comments on the "amazingly successful invasion of Old World livestock" in the Caribbean, projecting an easy adaptation to the new environment in which they thrived, with some becoming feral.[2] Noting the animal needs of early Caribbean plantations, David Watts discusses the heavy reliance on horses, donkeys, and cattle, as well as the attempt to use camels and the introduction of sheep, pigs, and other domesticated animals.[3] While Crosby argues that "many kinds of livestock pathogens have lagged behind their hosts in the trans-oceanic crossing,"[4] some arriving in the mid-nineteenth century, Watts refers to the vulnerability of horses and camels in Barbados to disease during the 1660s.[5] Indeed, we find one Trinidad planter complaining about a disastrous epidemic affecting mules on his estate in 1853.[6] From the 1930s, a body of literature on outbreaks of rabies in Trinidad was published in medical journals by Dr. Joseph Pawan, who was engaged in research on this disease.[7] Subsequent to this, veterinarians Harry Metivier and Holman Williams have written on the subject.[8] Metivier outlines the development of veterinary services in Trinidad and Tobago, while Williams offers a brief discussion on the origins of animals and their diseases in the region. Because of sporadic and limited treatment in the historiography, it is not possible to obtain a comprehensive picture of the nature of the impact of animals on the history of the British Caribbean. This chapter is an attempt to address this deficiency.

The chapter examines the impact of animals on the history of Trinidad and Tobago from 1879 to 1962 with specific reference to animal disease. The experience of this colony reflects an interrelationship between human and animal health at critical times in the historical evolution of the society. The chapter aims to demonstrate the significance of animal-centered research for Caribbean history with the hope that further research in this area will be stimulated. The focus of the chapter is first on the zoonoses that presented serious public health challenges in the colony. The high incidence of tuberculosis during the first half of the twentieth century and of the mystifying disease, which was later identified as paralytic rabies, reflected the intersection between human and animal health. The second focus is to show how animal diseases impacted the economy of the colony and consequently led to the establishment and development of veterinary services in Trinidad and Tobago from the late nineteenth century to the end of the colonial era.

The History of the Veterinary Establishment in Trinidad and Tobago

Veterinary concerns received growing focus during the second part of the nineteenth century. Increased sanitary consciousness in the late nineteenth century resulted from new knowledge on disease-causing agents and led to a growing emphasis on disease prevention at the imperial centers, which was slowly reflected in the policies in the colonies.[9] In particular, there was recognition of the role of the movement of people, animals, and plants in the spread of disease, and it is to this movement that the first colonial attempts at disease control were directed. Hence, quarantine was the first measure of disease control. The first specific animal-related action in this regard was the 1860 Trinidad ordinance for the control of the bacterial disease glanders, (or farcy) in horses. This law authorized the appointment of a veterinary officer to examine imported animals and destroy those that were infected.[10] Ordinance 19 of 1872 was also intended to prevent glanders.[11] This law restated the authority to appoint a veterinary officer and stipulated fines of fifty pounds for individuals in possession of diseased animals.

In keeping with the new thrust toward cleanliness was the attempt to ensure that clean meat was offered for sale to the public. Compulsory inspection of animals to be slaughtered by a competent veterinary officer was introduced in 1870 to ensure that diseases were not spread through the sale of diseased meat.[12] These precautions were deemed necessary because the annual importation of animals was of significance to the economy and society. Horses were imported for the police service, for transporting the upper class, and for upper-class sport, especially hunting and horse racing. Other animal imports included dogs for the police service and as pets, beasts of burden—bison, donkeys, and mules—and animals for the meat and dairy industry (see table 9.1). As table 9.2 demonstrates, imports continued to be a feature of life in the colony in the twentieth century. These included horses, mules, and donkeys that were imported from the United States, the United Kingdom, Canada, Venezuela, Barbados, St. Vincent, Demerara, and other British Caribbean colonies.

Trinidad was dependent on animal imports for food, transport, and breeding purposes. During the last quarter of the nineteenth century, there was a concerted effort to improve the milk- and meat-producing stock in

Table 9.1. Animal imports into the colony of Trinidad and Tobago, 1892–98

Year	Number of animals
1892	950
1893	1,419
1894	1,236
1895	1,286
1896	1,161
1897	1,077
1898	1,069

Source: CO 318/200, *Annual Report of the Government Veterinary Officer and Examiner of Animals of Trinidad and Tobago for 1898.*

Table 9.2. Animal imports into the colony of Trinidad and Tobago for the year 1933

Animal type	Numbers
Cattle	7,539
Horses	85
Mules	214
Donkeys	64
Dogs	57
Pigs	3,866
Sheep	1,392
Goats	4,297
Lions	4
Tiger	1
Deer	2
Cats	3

Source: CP no. 84 of 1934, *Administrative Report of the Director of Agriculture of Trinidad and Tobago for the Year 1933.*

the island because there were concerns with the health consequences of impure milk and tainted meat consumption across the British Empire.[13] The government took a leading role in this activity by establishing government stock farms that became the breeding centers for the sale of desirable stock, for diffusion of information, and for provision of related services. Animal breeding and disease prevention, therefore, constituted two essential elements of agricultural policy from the late nineteenth century.

The earliest initiatives did not result in the establishment of a veterinary system in the colony but remained symbols of a growing concern of the Central Board of Health, under whose ambit fell all health-related matters. In fact, the first quarantine officer was a medical officer, called the port health officer, who dealt with all health matters in the port. Metivier, in his account, dates the history of veterinary services in Trinidad and Tobago from the 1879 establishment of a stock farm to supply milk to the colony's hospitals, asylums, and other public institutions.[14] This government-operated scheme was a cost-cutting measure to provide a cheap supply of milk to government institutions and set high standards of milk quality. It was first essential to improve the colony's stock by importing desirable breeds. The farm served as a breeding station and sold pure-bred animals for milk, meat, and haulage. The farm was managed by the government veterinary surgeon and superintendent of government pastures; the first was J. B. White, who served from 1894 to 1895.[15] White was followed by F. Pogson, who served from 1898 to 1902, and then by D. Millar from 1902 to 1916. In 1916, Trinidadian Charles de Boissiere served as acting veterinary officer until 1920 when he went into private practice in Port of Spain.[16] In 1908, the Department of Agriculture was established to direct the colony's agricultural development. This department later assumed control of the government stock farms and the veterinary section.[17]

Metivier argues that the foundations of veterinary husbandry were laid by his father, Dr. Harry Metivier, who managed the government stock farm from 1916 and served as director of agriculture as well as a lecturer at the Imperial College of Tropical Agriculture. He also served as chief veterinary officer to the Mounted Police and advised them on the purchase of horses. In addition, he undertook research on rabies, for which he was honored by the French government with the Merite d'agricole.[18] In 1920, the veterinary establishment began to expand as sugar estates also began hiring veterinary officers. Usine Ste. Madeleine hired E. McClachlan to care for their draft animals—horses, mules, zebu, and water buffalo—and a small dairy herd that was used to supply milk to company staff. The company also employed J. Shannon until 1923 when he joined the Department of Agriculture.

McClachlan, who served the company from 1920 to 1956, was instrumental in reducing the high incidence of tuberculosis in water buffalo on company estates. These animals were highly valued as draft animals, for the high butter content of their milk, and for their tender meat. The success of the dairy at Usine Ste. Madeleine stimulated a similar program on other sugar estates and influenced stock production in South Trinidad.[19]

Under Metivier, milk production increased from 1,250 gallons per month from 100 cows to 120 cows producing 3,000 gallons per month.[20] The price of milk in the colony was reduced, but there was also an increase in revenue from milk production. Revenue to the colony from the sale of milk increased from £2,250 to £4,000 and provided training for local dairymen. Tuberculin testing of cattle was made mandatory for all milk-producing cows, milk vendors had to be licensed, and cattle were sprayed at intervals against ticks. In 1923 the Veterinary Department was expanded and shortly afterward became preoccupied with bat-borne rabies.[21]

There was further expansion of the veterinary unit when, in 1935, an officer was posted at the government stock farm in Tobago. In 1936, the department was reorganized to become a part of the colonial service, and new posts were created.[22] Professional status was afforded to veterinary officers in the colony by the 1930 Veterinary Surgeons' Registration Act, which came into effect on 2 June 1939;[23] and in 1934 government veterinary officers were allowed to have private practices under specified conditions.[24]

Regional consultations on animal matters began in 1947 when several colonies revised their regulations on animal health to adopt the British model. A Veterinary Subcommittee of the Official Standing Committee on Agriculture, Animal Health, Forestry and Fisheries helped to establish uniformity in the legislation of the departments in the region.[25] The work of the department was also greatly assisted by the establishment of a small Veterinary Diagnostic Laboratory at the St. Joseph Government Farm in June 1949.[26] This laboratory was administered by the veterinary officer at the government farm; its main areas of work were milk testing, pathological investigations, and equine-pregnancy diagnosis. These were later expanded to include poultry autopsies and serological tests.[27] In 1954 the laboratory was removed to the Central Experimental Station at Centeno.[28]

The Department of Agriculture was reorganized in 1958, and a program for agricultural development was outlined. The minister of agriculture assumed responsibility for shaping agricultural policy and an Animal Health Division was created within the Ministry of Agriculture. Its objectives were to implement animal-health regulations, undertake preventive medicine, and provide clinical, diagnostic, and investigative services.[29] By 1958, the veterinary establishment had finally taken shape and was positioned to play a vital role in the country's economy.

Several factors are noteworthy in the development of veterinary services in Trinidad and Tobago. In the first instance, this was the first government-controlled administrative unit to be established in an era when agricultural decision making was viewed as the exclusive domain of the individual planta-tion owners. Second, its activities were supported and augmented when sugar and, later, oil companies also established dairy herds and appointed veterinary officers, in their attempts to provide their staff with clean milk. Although its first stimulus was related to disease control, the development of this unit was closely related to the need to cut government costs and improve the quality and quantity of milk sold in the colony. The concern with wholesome milk was also representative of a trend that was visible in other territories of the region.[30]

The veterinary unit also represents the attempt to diversify agriculture in a colony that had hitherto focused almost exclusively on crop production. The significance of this development will be appreciated when it is viewed against the background of a beleaguered sugar industry of the late nine-teenth century and the near-collapse of the cocoa industry in the 1920s. It must also be mentioned that the planter community had resisted all previ-ous attempts to institute diversification strategies. Thus, the initiation and development of the veterinary unit must be seen as breaking new ground in the agricultural sector of the colony. At the same time, it must also be recognized that this activity allowed an administrative focus on an area of primary concern to the plantation owners and thus, inter alia, helped to strengthen the elitist element of the colony's agriculture. The discussion now turns to matters that preoccupied the veterinary establishment in the colony.

Veterinary Concerns in Trinidad and Tobago since the Late Nineteenth Century

Tuberculosis (Mycobacterium bovis)

The first concerns about this bacterial and highly infectious disease were related to its introduction into the colony from outside sources, which led to the examination of imported animals at the ports. Ships transporting animals would be quarantined if they had visited an infected port or con-veyed a sick or dead animal. Animals for local sale had to be examined, and the conditions for slaughtering and sale at the abattoirs and markets of the colony were monitored. There was legislation to prevent the spread of this and other diseases through the sale of infected meat and to improve the sanitation of public places.

Toward the close of the nineteenth century, the government veterinary officer noted the high incidence of bovine tuberculosis in animals in the colony.[31] Apart from bovine mortality, there was concern with the increasing number of human cases of tuberculosis in the colony. That this very infectious disease was a major cause of death and that it seriously affected the labor supply of the colony were causes for concern in themselves. The matter assumed greater urgency as it became known that this disease could be transferred from animals to humans. The milk supply, therefore, became important both as a source of infection and prevention. Regulations were established for the sale of milk. Small operators were accused of selling bad quality milk, which was often diluted and infected. The government assumed responsibility for effecting change in this area through the government farms, which served as the models for dairy and livestock operations in the colony.

The disease assumed greater significance for the colony's economy when increased emphasis was placed on livestock production. As the impact of the worldwide depression enveloped the major export crops, more attention was given to animal husbandry. Concerns about bovine tuberculosis spawned activity in three main areas. First, there was vigilance with respect to animal diseases in general. This entailed intelligence on the existence of diseases in countries with which Trinidad and Tobago maintained contact and the institution of careful port procedures and accurate diagnosis of each case of animal disease. This process entailed efforts to identify specific animal diseases, trace their origins, and keep details of their occurrence. As a result of this activity, other diseases such as Bulbar Paralysis and tick-borne red water fever or Bovine Piroplasmosis (*Babesia bigemina*) were recognized. This latter disease, caused by a parasitic protozoon, was associated with the large number of pure-breed cattle imported from Canada.[32]

Second, activities were specifically related to tuberculosis itself. In 1927 a committee to report on the prevalence of bovine tuberculosis was established, and in 1928 an amendment to the Disease of Animal Ordinance to secure the examination and testing of all cattle for bovine tuberculosis was passed.[33] There were reports of opposition to these new regulations. It was noted in 1947 that there was a high incidence of this disease in water buffalo on the sugar estates.[34] Eradication was made difficult by the fact that water buffalo showed little signs of debilitation even during the most advanced stages of the disease.[35] The sugar companies were urged to slaughter infected animals, but, citing concerns about costs, often they did not comply.[36]

The third area of activity was related to increasing and sanitizing the production of milk, which was an important component of the new thrust into livestock production. New regulations for tuberculin tests on animals were issued in 1928.[37] To ensure the colony of a copious supply of milk, pure-breed Holstein cattle were imported. Attention was given to the environment in which milk was produced. A new law required that dairy workers provide a medical certificate of good health to establish their fitness for employment in the dairy industry. A milk inspector was appointed on 1 February 1929, to ensure that the regulations were strictly implemented.[38] The veterinary laboratory moved into routine examination of milk produced on the St. Joseph Stock Farm and by peasants.[39] Testing for tuberculosis also became routine in the colony. Tuberculin tests were enforced in Port of Spain in 1933. It was mandatory for all dairy cattle supplying milk to the city to be tested. Reactors were slaughtered, and the government paid partial compensation to the owners.[40] But there continued to be a high incidence of bovine tuberculosis even as the initial resistance to testing dairy cattle was reduced.

The issue of tuberculosis was addressed at the Caribbean Livestock Conference in 1951, which recommended that the Bureau of Animal Industry (BAI) tuberculin testing be used. This recommendation was accepted in Trinidad and Tobago, where it was also decided that all future testing should be conducted by government veterinary officers. The mandatory certificates for the sale of milk would be issued only if such tests were carried out.[41] While some progress was made with reducing the occurrence of tuberculosis, another major animal disease appeared on the landscape.

Paralytic Rabies

Between 1925 and 1931, a number of domestic animals, particularly cattle, succumbed to a "strange" disease that was believed to be either botulism or a form of polio. Cane and cocoa farmers lost over five hundred head of cattle in 1932.[42] The government set up a special committee made up of the director of agriculture, two veterinary officers, and the government pathologist to investigate the disease. The committee reported in 1932 that the disease, paralytic rabies, was caused by a virus and was spread by vampire bats, especially the species *Desmodus rotundus*.[43] A mass inoculation drive against rabies commenced in 1932. Estates were charged one shilling a dose for the inoculation, while the service was free to peasants. Over six thousand animals were inoculated in 1932 when emergency measures were implemented.[44]

There was a major outbreak of paralytic rabies in the colony in 1935 when there were 331 cases in cattle and twenty-one reported human cases.[45] An anti-bat program, to capture and destroy *Desmodus rotundus*, was launched in the colony.[46] The death rate from rabies was very high, and over 90 percent of the cases occurred in cattle.[47] Despite an eight-year lull in the incidence of the disease in humans and a reduced incidence in animals, the inoculation and the anti-bat programs were intensified in 1948.[48] Sporadic outbreaks of the disease kept the anti-rabies unit busy. There was a small outbreak in Salazar Trace, Point Fortin, in 1951,[49] and cases of rabies occurred in North and South Trinidad in 1952, 1953, and 1954. Both the Departments of Health and Agriculture mounted programs to deal with this disease. The Health Department continued the anti-rabies program, which included the search for and destruction of *Desmodus* bats and their roosts and inoculation to humans. The Department of Agriculture carried out protective inoculation of livestock in the infected areas.[50]

As a result of investigations into the 1954 epidemic in South Trinidad, the mongoose was examined as a vector of rabies.[51] The recurrence of the disease in the decade of the 1950s led to intensification of the activities of the anti-rabies unit.[52] There were forty-two cases of rabies, and 4,505 animals were vaccinated in 1953.[53] The situation assumed crisis proportions when an epizootic occurred in the south of Trinidad in 1954.[54] Assistance was solicited from the World Health Organization (WHO).[55] The Departments of Health and Agriculture continued their attempts at mass inoculation and intensified bat destruction, but the opposition of a large number of farmers, who continued to discredit vaccination and refused the free service until it was too late, presented a serious obstacle. The departments then began an education campaign to make farmers more vaccination-conscious.[56] The following table reflects the increasing numbers of vaccinations and the cases that were reported for each year between 1955 and 1957.

As a part of the anti-rabies campaign, emphasis was placed on further training of officers and an improved administration. A Standing Paralytic Rabies Committee—made up of representatives of the Department of

Table 9.3. Paralytic rabies: Vaccinations and cases, 1955–57

Year	Vaccinations	Cases
1955	9,780	22
1956	24,751	3
1957	25,921	3

Source: *Administrative Report of the Director of Agriculture for the Year 1957*, 45.

Health, Trinidad; Regional Virus Laboratory; the medical officer of health, Port of Spain; veterinary officer (rabies); and the deputy director of agriculture (animal husbandry)—was established in 1956.[57] With WHO assistance, the committee formulated a comprehensive program for rabies control.[58] Provision was made, through WHO fellowships, for the training of veterinary officers in rabies control, so they could assume responsibility for all related field and laboratory work. The division received additional staff, and the cattle immunization program was continued. Control of field operations of bat control, which were formerly handled by the Department of Health, was allocated to the Department of Agriculture in 1956.[59] The final segment of the rabies-control program was the implementation of mass inoculation of cattle and of people involved in the campaign.[60] Hence, while there were eighteen cases of rabies in 1958, there were 39,733 inoculations, and 1,546 vampire bats were destroyed. In 1962, there was only one recorded case of rabies, but 27,950 animals were vaccinated and 1,466 bats destroyed.[61]

It is clear that efforts to combat paralytic rabies came to dominate the energies of both the health and agricultural departments of Trinidad and Tobago between the 1930s and 1960. The economic impact of this disease must not only be assessed in terms of the value of the animals lost but must also be seen in terms of the costs to the colony of the vaccination and antibat campaigns, both of which experienced considerable expansion across the period. The efforts to combat the disease stimulated significant local research. Trinidad became noted for the classic epidemiological studies of bat-transmitted rabies, which were conducted by government bacteriologist Dr. Joseph Pawan. Research and laboratory testing and analysis helped to save the livestock industry of Trinidad and Tobago through the critical era of the ravages of rabies. The Trinidad situation was further complicated by the eruption of disease in other sectors of the livestock industry.

Poultry Diseases

Poultry diseases caused significant losses to those engaged in poultry production during the second half of the twentieth century. As a significant poultry industry developed in the colony, one of the first diseases to appear was fowl cholera. This virus caused fifty-four outbreaks of the disease, which caused serious losses across the country in 1951.[62] The first outbreak of Newcastle disease (avian pneumoencephalitis), another virally transmitted disease, occurred in 1952.[63] It was reported that these occurrences of poultry diseases adversely affected the local supply of meat and eggs and the incomes of numbers of small poultry farmers.[64]

There was a recurrence of both diseases in 1954, when thirty-two out-breaks of fowl cholera and nine outbreaks of Newcastle disease were re-ported.[65] Considered a serious threat to the poultry industry, Newcastle disease was then made notifiable, and an intranasal vaccine was used to immunize chickens. The disease was fought by an information campaign to educate farmers on its symptoms and treatment and by the institution of measures for the vaccination of all chicks before delivery.[66]

Newcastle disease reappeared in 1955 when there were twenty-two outbreaks,[67] along with eight outbreaks of fowl typhoid and again in 1957 when three outbreaks were reported.[68] It was, however, the outbreak of Newcastle disease in 1960–61 that was most devastating for the colony's poultry industry. With reports of twenty-one outbreaks, there were over two hundred thousand fatalities.[69] On large farms, tractors were used for mass burials, and the government spent $189,534 EC to provide loans to the thirty-four farmers affected.[70] After this major outbreak and as vaccination became more common, the occurrence of Newcastle disease declined in Trinidad and Tobago.[71] The authorities were also required to give their attention to porcine diseases.

Classical Swine Fever

Pig rearing was a popular activity among the freed population, but there are no records of large-scale disease problems until the 1940s. According to Metivier, one impact of the American presence at the bases in Trinidad during World War II was the introduction of swine fever into the colony. He states that with the post-1940 influx of American soldiers came the introduction of infected American meat. The first outbreak of this disease in the colony occurred in 1946. Its introduction was traced to swill, which contained infected American meat that was fed to local pigs.[72] There were thirty-two reported cases of swine fever in this outbreak. Vaccination with crystal violet vaccine was regularly carried out at the government stock farms, especially on young pigs as soon as they were weaned.[73] The vaccine was made available to the farming community at the cost of forty-eight cents a dose.[74]

Continued outbreaks, thirteen in 1950, reduced the domestic supply of pork and hampered the movement of pigs.[75] In 1952, 150 pigs died, and forty-seven were slaughtered in the swine-fever outbreak. In that year, 1,206 pigs were vaccinated. Further losses were also incurred from a disease identified as necrotic enteritis (*Clostridium perfringens*, Type C).[76] There were no reported cases of swine fever between 1953 and 1957, but the vaccination program was maintained. In 1953, 1,458 pigs were vaccinated,[77] 1,876 in 1955,[78] 1,892 in 1956,[79] and 1,659 in 1957.[80]

The most widespread outbreak of swine fever in Trinidad and Tobago occurred in 1960 when it is estimated that over thirty-five hundred pigs died.[81] Losses from this outbreak were assessed as $140,000, exclusive of vaccines, labor, and other inputs. Mortality was 100 percent. This epizootic demonstrated how relatively unvaccinated the swine population of Trinidad and Tobago was. In 1959, 2,098 of an estimated swine population of over thirty thousand were vaccinated. As a result of the outbreak, over twenty-six thousand pigs were vaccinated in 1960 with inactivated tissue vaccine or antiserum. There was a recurrence of this disease in 1962 when 49 swine fever outbreaks signaled the start of another epizootic which continued into 1963.[82] While 7,927 pigs were vaccinated during the year, it was clear that the inoculation program was not adequate.[83] The problem was that no clear policy on swine fever had been developed in the colony.[84] The response to the vaccination program was poor, and the aggressive drive evident in the programs against tuberculosis and rabies was not evident. At the government farms, piglets were vaccinated before sale, but this example was not commonly followed by members of the farming community. The result was that, at the end of the colonial period, the issue of swine-fever infection remained a threat to pig rearing in Trinidad and Tobago. Unlike cattle rearing, pig rearing was an activity that was of major importance to small farmers. The impact of this disease on poor families was, therefore, significant.

THIS DISCUSSION of the history and development of veterinary services in Trinidad and Tobago between 1879 and 1962 has demonstrated that these services originated from two main developments in the colony. First, there was the conviction that agricultural diversification was the way forward for the British Caribbean territories during the latter part of the nineteenth century. At the same time, the need of the colonial government to reduce its overhead led to a focus on milk production. Government interest in the dairy industry was further justified by the claims that the quality of milk sold in the colony was poor and that small operators and vendors did not pay adequate attention to sanitation. This provided the impetus for state intervention, through the establishment of an experimental farm, to stimulate a dairy industry and livestock production in the colony. While this farm produced milk to supply government institutions, it was responsible for importation of desired breeds, conducting breeding experiments, providing services and information to livestock farmers, and also serving as a model operation for farmers. In this way the state had taken initiative in promoting diversification in the colony.

Second, veterinary services also developed as a part of the colony's response to animal disease. While there were incidences of diseases in animals since the early colonial period, these were dismissed as "strange" occurrences and have not been examined in the historiography of this colony. When, from the late nineteenth century, efforts were made to diversify the economy of the islands, both the number and variety of breeds of domestic animals were increased. Thus, the composition of the animal kingdom in the colony was changed. As a result of increased imports, there was an increase in the incidence of animal disease in the colony. The presence of these diseases changed the disease environment of the islands and threatened the very economy that the animals were brought to improve.

There are two factors that are noteworthy in the development of veterinary services in Trinidad and Tobago. There was the strong research element that characterized veterinary activities. This research effort, which traced the origins of disease, was the critical factor in the efforts to eradicate animal diseases in the colony. The second factor is that the veterinary division was the first agricultural section to be provided with a formal public administrative structure. Hence, this division predated the Department of Agriculture of which it would later become a part. With an organized structure, it was this division that was first poised to assist the economy of the colony to move into greater diversification in an era when the main agricultural interests remained primarily concerned with the established plant cultivations. This was an important contribution to the economic development of the colony. With its initial focus on the animal interests of the upper class, the veterinary division helped to maintain the existing class divisions in agriculture.

During the course of the period under discussion, veterinary services came under the ambit of both the Departments of Health and Agriculture. The efforts of the veterinary officers were initially focused on the animals that were of greater interest to the upper classes—horses and cows. Later, the poultry industry was drawn into its ambit. At the end of the selected period, an effective system for dealing with the diseases of pigs had not been formulated in the colony. Thus, across the period, the small farming community suffered severe losses from the ravages of the diseases of pigs. At the end of the period and in response particularly to the problems associated with the outbreaks of rabies, the Animal Health Division was fully established as a part of the Department of Agriculture and its services were made more generally available to the farming community. Thus, the institution of veterinary services in Trinidad and Tobago was an initiative of the imperial government in its efforts to diversify and strengthen the economy of the colony. The operations of the veterinary establishment in Trinidad and Tobago demonstrate the important ways animals have had an impact on the history of the Caribbean and underscore the need for closer examination of their role in the evolution of Caribbean societies.

Notes

1. Holman E. Williams, "Caribbean Community (CARICOM) Food Security-Livestock Diseases" (paper presented at the Fifth Regional Livestock Meeting, Regional Livestock Development Programme, Bahamas, 26–28 September 1984), 2. A discussion of the domesticated animals introduced in the West Indies by Europeans is provided in David Watts, *The West Indies: Patterns of Development, Culture and Environmental Change Since 1492* (Cambridge: Cambridge University Press, 1987), 162–64.

2. Alfred Crosby, *The Columbian Exchange: Biological and Cultural Consequences of 1492* (Westport, CT: Greenwood, 1972), 77.

3. Watts, *West Indies*, 197–99.

4. Alfred Crosby, *Ecological Imperialism: The Biological Expansion of Europe, 900–1900* (Cambridge: Cambridge University Press, 1986).

5. Watts, *West Indies*, 198.

6. *San Fernando Gazette*, 6 October, 1854.

7. See the following four articles: E. W. Hurst and Joseph L. Pawan, "An Outbreak of Rabies in Trinidad without History of Bites and with the Symptoms of Acute Ascending Myelitis," *Caribbean Medical Journal* 21, nos. 1–4 (1959): 11–24; Hurst and Pawan, "And Further Account of the Trinidad Outbreak of Acute Rabic Myelitis: Histology of the Experimental Disease," reprinted from *Annals of Tropical Medicine and Parasitology* 30, no. 4 (1936), in *Caribbean Medical Journal* 21, nos. 1–4 (1959): 137–65; J. L. Pawan, "The Transmission of Paralytic Rabies in Trinidad by the Vampire Bat (Desmodus Rotundus Murinus Wagner, 1840)," *Caribbean Medical Journal*, 21, nos. 1–4 (1959): 110–34; J. L. Pawan, "Rabies in the Vampire Bat of Trinidad with Special Reference to the Clinical Course and the Latency of Infection," *Caribbean Medical Journal* 21, nos. 1–4 (1959): 25–45.

8. Hugh N. Metivier, *A History of Overseas Veterinary Services, Part II: Trinidad and Tobago, 1879–1958* (London: British Overseas Veterinary Association, 1973); Holman E. Williams, "The Genesis and Dynamics of Animals and Their Diseases in the Caribbean (in Parallel with Human Endeavour)," *Journal of the Caribbean Veterinary Medicine Association* 2, no. 1 (2002): 15. I am indebted to veterinarians Dr. Steve Bennett, Professor Holman Williams, and Dr. Joseph Ryan, who provided valuable information and assistance to this research project.

9. See, for example, Juanita de Barros, *Order and Place in a Colonial City: Patterns of Struggle and Resistance in Georgetown British Guiana* (Montreal: McGill-Queens University Press, 2002).

10. Ordinance 15, 23 June 1860, Trinidad Ordinances, PRO, CO 297/8.

11. Ordinance 19 (Ordinance for the Prevention of Glanders), Trinidad Ordinances, 1 November 1872, PRO, CO 297/9.

12. Trinidad: Minutes of the Legislative Council, 10 December 1870, PRO, CO 298/35.

13. De Barros, *Order and Place*, 122–31.

14. Metivier, *History of Overseas Veterinary Services*, 329.

15. Ibid.

16. Ibid.

17. Rita Pemberton, "The Evolution of Agricultural Policy in Trinidad and Tobago, 1890–1945" (PhD thesis., University of the West Indies 1996), 105.

18. Metivier, *History of Overseas Veterinary Services*, 329–30.

19. Ibid., 330–31.

20. Ibid., 331.

21. Ibid.

22. Council Paper (CP) no. 47 of 1937, *Report of the Veterinary Division for 1936*, 49.

23. Williams, "Genesis," 15.

24. *Administration Report of the Director of Agriculture for 1953* (Port of Spain: Government Printing Office, 1954), 38.

25. Holman Williams, "Animal Health Limitations within the Caribbean" (paper presented at Seminar on the Utilization of Local Ingredients in Animal Feedingstuffs, Kingston, Jamaica, 1975), 4.

26. *Administration Report of the Director of Agriculture for the Year 1953*, 17.

27. Ibid., 38.

28. *Administration Report of the Director of Agriculture for the Year 1954* (Port of Spain: Government Printing Office, 1955), 52.

29. *Administration Report of the Director of Agriculture for the Year 1958*, 34, in West Indiana Collection, University of the West Indies (UWI), St. Augustine.

30. De Barros, *Order and Place*, 131–37.

31. CP no. 63 of 1897, *Annual Report of the Government Veterinary Surgeon and Examiner of Animals for the Year 1896* (Port of Spain: Government Printing Office, 1897), 3.

32. CP no. 54 of 1930, *Administration Report of the Director of Agriculture for the Year 1929* (Port of Spain: Government Printing Office, 1930), 21.

33. CP no. 63 of 1930, *Medical: Administrative Report of the Surgeon General for the Year 1929* (Port of Spain: Government Printing Office, 1930), 4.

34. *Administration Report of the Director of Agriculture for the Year 1949* (Port of Spain: Government Printing Office, 1951), 18.

35. *Administration Report of the Director of Agriculture for the Year 1952* (Port of Spain: Government Printing Office, 1954), 43.

36. *Administration Report of the Director of Agriculture for the Year 1950* (Port of Spain: Government Printing Office, 1952), 22.

37. CP no. 54 of 1930, *Administration Report of the Director of Agriculture for the Year 1930*, 20.

38. *Administration Report of the Director of Agriculture for the Year 1951* (Port of Spain: Government Printing Office, 1953), 21.

39. Ibid.

40. *Administration Report of the Director of Agriculture for the Year 1947* (Port of Spain: Government Printing Office, 1948), 4.

41. *Administration Report of the Director of Agriculture for the Year 1950* (Port of Spain: Government Printing Office, 1952), 22.

42. CP no. 33 of 1933, *Administration Report of the Director of Agriculture for the Year 1932*, 12, 39.

43. Ibid. The Trinidad vampire bat was initially identified as *Desmodus rufus* Wied, 1826, but the specific name is *Desmodus rotundus murinus* Wagner, 1840. See J. L. Pawan, "Transmission," 114.

44. *Administration Report of the Director of Agriculture for the Year 1932*, 12, 39.

45. CP no. 47 of 1937, *Report of the Veterinary Division for the Year1936*, 50.

46. CP. no. 94 of 1936, *Report of the Surgeon General for the Year 1935*, 7.

47. CP no. 84 of 1934, *Administration Report of the Director of Agriculture for the Year 1933*, 30.

48. *Medical and Sanitary Report of the Director of Medical Services, Trinidad and Tobago for the Year 1948*, 6, in West Indiana Collection, UWI, St. Augustine, Trinidad and Tobago.

49. *Medical and Sanitary Report of the Director of Medical Services, Trinidad and Tobago for the Year 1951*, 7, in West Indiana Collection, UWI, St. Augustine, Trinidad and Tobago.

50. *Medical and Sanitary Report of the Director of Medical Services for Trinidad and Tobago for the Year 1952*, 6; *Report of the Director of Medical Services for the Year 1953*, 7; *Medical and Sanitary Report of the Director of Medical Services Trinidad and Tobago for the Year 1954*, 8, in West Indiana Collection, UWI, St. Augustine, Trinidad and Tobago.

51. *Medical and Sanitary Report of the Director of Medical Services for Trinidad and Tobago for the Year 1955*, 7, West Indiana Collection, UWI, St. Augustine, Trinidad and Tobago.

52. *Trinidad and Tobago: Medical and Sanitary Report of the Director of Medical Services for the Year 1951*, 7.

53. *Administration Report of the Director of Agriculture for the Year 1953*, 36.

54. *Administration Report of the Director of Agriculture for the Year 1954*, 51.

55. Ibid.

56. Ibid., 51–52.

57. *Administration Report of the Director of Agriculture for the Year 1957*, 25.

58. *Trinidad and Tobago: Annual Report of the Department of Agriculture for the Year 1958*, 33.

59. Ibid.

60. Ibid., 35–36.

61. *Trinidad and Tobago: Annual Report on the Agriculture Services for the Year 1962* (Port of Spain: Government Printer, 1963), 5.

62. *Administration Report of the Director of Agriculture for the Year 1951*, 25.

63. Metivier, *History of Overseas Veterinary Services*, 336.

64. *Administration Report of the Director of Agriculture for the Year 1950*, 4.

65. *Administrative Report of the Director of Agriculture for the Year 1954*, 20.

66. *Administration Report of the Director of Agriculture for the Year 1952*, 42.

67. *Administration Report of the Director of Agriculture for the Year 1955* (Port of Spain: Government Printer, 1957), 36.

68. *Administration Report of the Director of Agriculture for the Year 1957*, 45.

69. *Administration Report of the Director of Agriculture for the Year 1960* (Port of Spain: Government Printer, 1965), 23.

70. Williams, "Animal Health Limitations within the Region," 1.

71. *Administration Report of the Director of Agriculture for the Year 1956* (Port of Spain: Government Printer, 1958), 37.

72. Metivier, *History of Overseas Veterinary Services*, 336.

73. *Administration Report of the Director of Agriculture for the Year 1947* (Port of Spain: Government Printer, 1948), 17.

74. *Administration Report of the Director of Agriculture for the Year 1949*, 17.

75. *Administration Report of the Director of Agriculture for the Year 1950*, 4.

76. *Administration Report of the Director of Agriculture for the Year 1952*, 44.

77. *Administration Report of the Director of Agriculture for the Year 1953*, 37.

78. *Administration Report of the Director of Agriculture for the Year 1955*, 33.

79. *Administration Report of the Director of Agriculture for the Year 1956*, 37.

80. *Administration Report of the Director of Agriculture for the Year 1957*, 26.

81. *Administration Report of the Department of Agriculture for the Year 1960*, 22.

82. *Trinidad and Tobago: Annual Report of the Department of Agriculture for the Year 1963* (Port of Spain: Government Printer, 1964), 6.

83. *Trinidad and Tobago: Annual Report of the Department of Agriculture for the Year 1962* (Port of Spain: Government Printer, 1964), 6

84. Ibid.

CHAPTER 10

Nineteenth-Century Australian Pastoralists and the Origins of State Veterinary Services

John Fisher

IN 1993, Sylvie Lepage spoke to a French couple who had bought a ten-thousand-acre property at Cargo in central New South Wales in 1983. Asked why two Parisians became sheep farmers in Australia, Frederic explained:

> Space and climate. You don't have to grow fodder here unless there is a drought. The animals stay outdoors all year round. The only building you need is a shearing shed. We don't have veterinary expenses either, whereas in Europe diseases become more contagious because animals are packed closer together. They have to be looked after. Here we practise natural selection.[1]

The contribution of space to the success of livestock production in Australia has long been clear, but the low cost of veterinary care also played a major role from the beginning of white settlement. At first this was because most of the infections and infestations that livestock suffered from in Europe found it difficult to survive in the distinctive ecosystems of Australia. As this freedom from disease did not persist, however, action to control introduced diseases and to prevent further imports was necessary.

The evolving response to disease led to the formation of organizations that came to be called "stock branches" in the Australian colonies. They appeared soon after the middle of the nineteenth century, roughly contemporaneous with the appearance of state veterinary services in western Europe and other regions of European settlement and performing similar functions.[2] They differed, however, from the European model in two important respects. In the first place, although the stock branches were government bodies, with their powers established by legislation, their policies and actions were paid for and largely decided by their clients, the livestock owners themselves. In the second place, while in Europe the new services were staffed by professional veterinarians, the latter were markedly absent from the Australian stock branches.

These features arose out of the nature of Australian pastoralism as it evolved after 1788. The response to disease was one part of the evolving strategies of pastoralists as they sought to adapt Old-World technology and modes of production to meet the challenges that came from operating in the unfamiliar ecosystems of a New World. Their ability to adapt successfully, as will be seen below, reflected the particular characteristics of a distinctive group of large capitalist producers.

Evolution of Australian Pastoralism

The first white settlement in Australia was intended to be a jail, and this remained its primary official function until the 1820s. Nevertheless, many of the early settlers, notably John Macarthur, were attentive to the potential for economic gain.[3] Opportunities came first in supplying the jail at Sydney Cove with basic necessities (including rum), in facilitating Pacific trade, and in harvesting the marine resources (primarily whales and seals) of the South Pacific. It was soon evident, however, that the longer-run economic future of New South Wales lay in utilizing its most abundant factor—land—for livestock production.

Colonial governors and private individuals both played a role in the early introduction of domesticated livestock into New South Wales, but it was the latter who dominated the development of Australian pastoralism. After initial problems in getting horses, cattle, and sheep to Australia, the white settlers were encouraged by positive feedback from the processes Alfred Crosby terms "ecological imperialism."[4] Livestock flourished in Australian ecosystems in the early colony; they benefited from a relative absence of predators and disease agents, exhibiting high reproduction rates and excellent health. The problem for white settlers was how to extract maximum value from this success.

Despite a sharp rise in convict numbers after 1815, the jail was too limited a market to provide much of a prospect for colonial entrepreneurs. A number of them, however, were already experimenting with sheep breeding. John Macarthur, long considered the "Father of the Australian Wool Industry," was, in fact, only one of a number of entrepreneurs involved in the early development of Merino wool production.[5] However, the point here is that the successful pioneers were men and women with capital and vision combined with the ability to participate in the wider economic and social networks necessary to realize the fruits of their efforts.

Their success also reflected their capacity to operate on a large scale of production. Although there were numerous small farmers in the white settlement, mainly ex-convicts, these made little contribution to its growth. Scale of production became even more of an imperative when, from the 1820s, it became evident that the economic future of the colony lay in producing wool and a headlong process of spatial expansion began.[6] The adaptation of conventional British livestock husbandry to Australian conditions was already underway, with mixed farming giving way to extensive pastoralism, especially as stockowners moved into the grasslands to the south and west of the Sydney Basin and into the Hunter Valley. Spatial expansion brought with it the reduction in unit costs necessary if Australian producers were to compete in the British wool market and confirmed the trend toward large production units. This was accentuated by the environmental conditions that shaped the development of pastoralism as it spread through southeastern Australia.

Despite the advantages accruing from "ecological imperialism," expansion inland was as much a response to negative pressures as to opportunity. Sheep became increasingly "unthrifty" in the Sydney Basin, a result probably, according to Garran and White, of increasing internal worm infection.[7] "Unthriftiness" among all forms of livestock was also due to "coastal disease," a function of mineral deficiencies in local soils; by the 1850s there were few sheep left in coastal New South Wales.[8] The unpredictability of rainfall also put a premium on the size of pastoral operations: holding land in a variety of regions reduced risk as it allowed for a degree of stock movements between these. Such strategies evolved necessarily on the basis of trial-and-error learning and, while there was a large element of fortune to individual success, as individuals, Australian pastoralists had to be adaptable and innovative in meeting a range of environmental challenges. They could also be creative when acting as a group, as their response to the problem of introduced disease demonstrates.

Australian Pastoralists and Sheep Scab

As far as pastoralists were concerned, there were two major threats to the viability of sheep farming from the 1820s: the indigenous inhabitants and the introduced disease of sheep scab. Modern historians have been far more interested in Aboriginal resistance to pastoral expansion, in the context of race relations generally, than in the disease.[9] Nineteenth-century sheep owners, however, had no doubt which was the more serious threat. As Robert Dawson, the chief agent of the Australian Agricultural Company, asserted in 1832, it was "more virulent and troublesome to get rid of here than in any other country with which I am acquainted."[10] Much more capital and effort were expended on dealing with scab than on coping with the indigenous people. Further, it was the spur for a sequence of innovations, technical and institutional, the latter leading to the formation of the stock branch.

Sheep scab mites or acarids (*Psoroptes ovis*) came to Australia in the First Fleet; this was one microorganism that found it easy to surmount the usual barriers to introduction.[11] This was because the mites can survive for long periods away from their hosts, a factor, too, in the early spread of infestation in Australia. The disease spreads through animal-to-animal contact; in the early settlement and during the pastoral expansion, intermingling of flocks was difficult to avoid. There was a further problem. The bites of the mites can cause intense itching, loss of condition, wool deterioration and shedding, exhaustion, and even death. However, in the Old World, if detected early enough, there were sufficient remedies, such as arsenic and turpentine-based dressings, to keep scab mites from being more than a nuisance. Detection was the role of shepherds, "the aristocracy of the agricultural labour force" because of their skills in containing disease: "It is ever the shepherd's duty to try to discover the cause of ailing and to supply the remedy."[12] But Australian shepherds, overwhelmingly convicts in the first fifty years of settlement, lacked the skills necessary either to detect the mites or treat them; as flocks got larger and expansion proceeded apace, conditions were ideal for the unchecked spread of infestation.

The search for remedies for scab led pastoralists to experiment with a variety of acaricides while, most significantly, they also developed more cost-effective means of treating infested sheep. In the 1840s these led to the development of simple technical innovations, involving a combination of dips, races, and drafting gates, which enabled the treatment of thousands of sheep daily.[13] Before then, however, "the principal Graziers of the Colony" used their power in the recently established New South Wales Legislative

Assembly to attempt a regulatory solution. In 1832 the Scab Act was passed under which magistrates were given the power to impound infested sheep when found traveling or in markets.

This was a limited measure but, even so, aroused considerable controversy. The act's sponsors aroused suspicion; the colony's history was marked by antagonism between leading settlers and the majority of the white population. Further, as the governor put it, "It was found difficult to introduce any measure which appeared likely to prevent the spread of the disease without at the same time interfering intolerably with the right of private property."[14] The 1832 act was, however, hardly a flagrant breach of such rights. Under the common law, owners whose sheep had become scabbed had a right to damages from the perpetrator; the act was an attempt to use state power the better to enforce such damages. But it was important because it was the first in a series of measures that saw the pastoral interest deviate from British precedent in dealing with livestock diseases in a manner that did eventually infringe decidedly on property rights.

This was although it seemed unlikely that the act was a notable success. New South Wales was still a jail, but its police, who were supposed to enforce the act, were notoriously overstretched and incompetent. It was also unlikely that they were better able to recognize the disease, especially in its early stages, than were the colonial shepherds. Nevertheless, the act was renewed two years later and in subsequent years, the only modifications being to make penalties harsher and to extend its provisions to a new disease threat, the mysterious "catarrh," the cause of a large number of sheep deaths in the decade after 1835.[15]

The sequence of Scab Acts did coincide with a decline in scab incidence in the Settled Districts of New South Wales.[16] The disease was most prevalent on the pastoral frontier, in the regions south of the Murray River, soon to become the colony of Victoria, and on the Darling Downs to the north. These were also the areas where sheep numbers were highest by the 1850s, and the consequent price differentials meant that they began to be imported from Victoria, where scab was still virulent, into New South Wales. The growth of urban demand also meant that the two major livestock markets in Maitland and Sydney became centers for the dissemination of scab as store sheep were sold for fattening. A new series of outbreaks led the pastoral interest in New South Wales to build on its established regulatory approach, but employing much more radical features.

A select committee of the Legislative Assembly, dominated by leading pastoralists, was appointed in 1854, and its recommendations were made with a view to "stamping out" scab in New South Wales. All diseased and in-contact sheep were to be destroyed and the carcasses buried or burnt. This applied to infested runs as well as stock being moved or at market. The constabulary, directed by the magistrates, was still responsible for enforcement, although two inspectors were to be appointed to police intercolonial imports on the Murray and at Sydney. The committee also recognized the need to provide compensation to the owners of slaughtered sheep. This was to be set at four shillings a head (about three-quarters of prevailing average values), the cost to be met by a levy of two pounds per one thousand sheep on all owners of one hundred sheep or more.[17]

The recommendations were radical on two scores. On the first, slaughtering sheep on an owner's land entailed an extreme invasion of private property rights. The committee anticipated likely objections, citing European and even eighteenth-century British precedents (in programs to eradicate rinderpest) for such a measure and arguing that their proposal only extended private and voluntary insurance schemes of slaughter already in existence in some districts. The point was not made explicitly, but what they proposed was to eliminate the "free-rider" problem inherent in these by employing the authority of the state to make them compulsory and universal. On the second score, a levy was a necessity because the colonial government made it clear that it was not prepared to pay for such a measure. The proponents of the bill sought to soften the impact, first by exempting small flock owners and second by arguing, on the basis of some highly optimistic forecasting and arithmetic, that it would only be needed for one year.

There was opposition from some pastoralists, but a feature of the progress and enactment of the Scab and Catarrh Bill's incorporating the proposed measures was the general support it received. The chief spokesman for its proponents, Augustus Morris, the member for the Liverpool Plains, claimed that, on the basis of petitions and evidence to the committee, they represented the owners of a "total amounting to nearly one-half of the entire sheep of the colony."[18] Subsequent compliance was high among the thousand owners who came under the act, while an attempt at repeal a year later found little support.[19] A regulatory apparatus under which pastoralists paid for measures that could entail the forcible destruction of their own property was firmly in place.

Why were they willing to do this? The Scab and Catarrh Act of 1854 was striking testimony to the concern pastoralists felt over the threat posed by scab. It was not just a threat to their critical capital asset, although this gained force from the premium that had developed for New South Wales sheep over Victorian. In the 1840s, "clean sheep" were "worth at least 2/6d per head more than scabby ones" or even sheep suspected of being scabbed, a premium of some 25 percent. By the 1850s, this also applied against sheep that had been dipped, in part because of the reasonable suspicion that this was an inadequate safeguard, in part because fleeces lost quality through being dipped.[20] Further, sheep values underpinned the viability of transactions in land, a subject of intense speculation at this time. As it had been from the 1820s, pastoral expansion was still a process of speculation,[21] with many participants seeking to profit from selling land and "runs" (lands under occupational leasehold) as much as from selling wool.[22] Eliminating the threat posed by scab seemed worth the price of a temporary levy.

Action under the act was successful, but only in the context of a series of problems.[23] It was soon evident that the police force could not be used as an enforcement agency; an inspectorate had to be appointed—and paid—instead.[24] These were not professional veterinarians; use of the latter was never considered, and there were few in the colony anyway. The inspectors tended to be failed pastoralists, of whom there was an ample supply, men supposedly able to recognize infested sheep. Doubts were later raised over their competence, but the more immediate problem was that eradication took longer and cost more than had been envisaged.[25] The levy, paid into what became known as the "Scab in Sheep Fund," had to be maintained for three years. However, an even greater problem loomed, one that pastoralists were reluctant to confront: how was the colony's disease-free status to be maintained?

The necessity of monitoring intercolonial imports had been recognized in the act, and, after 1855 there was a total ban on movement across the Murray. However, while it was easy enough to proclaim quarantines and bans, there remained a considerable financial incentive to move sheep from Victoria, where scab was still ubiquitous, either to the Sydney meat market or as stores for restocking inland stations. Further, while border controls were maintained after 1857, the amount in the Scab Fund slowly diminished.[26] A select committee in 1858 recognized the problem but balked at reimposing the levy.

The next and inevitable major outbreak of scab came in 1863. It was anticipated—a drought in the south in 1862 led to an increase in sheep movements in search of fodder along the Murray and an increase in shipments to Sydney.[27] In response, some scabby sheep were slaughtered; more inspectors were hired but at the cost of exhausting the Scab Fund. As in 1854, the colonial government made it clear that pastoralists could have any measure they wanted—as long as they paid for it themselves. The latter were riven by divisions; a supposedly unanimous delegation to the government calling for the immediate destruction of infested sheep was reduced to incoherence when the financial implications were spelled out to them.[28]

Salvation came from outside the colony. Pastoralists in all colonies had been experimenting for some time with a variety of acaricides to use as dipping mixtures.[29] The problem, however, was that, while these killed the scab mites, they did not penetrate the fleece sufficiently to kill the eggs. Then, in the late 1850s, a Victorian grazier, John Rutherford, found that he could do this with a mixture of sulphuric acid and tobacco boiled to the point where it was "as hot as the men could stand."[30] Despite its effectiveness, Rutherford's innovation was slow to be adopted, probably because of its expense. Heating up masses of water required a substantial investment in tanks and boilers (although these were present on the largest runs, where pastoralists had responded to trade conditions by washing sheep before shearing).[31] Unknown in New South Wales until 1863, it then allowed pastoralists to evade the dilemma they had faced.

Alexander Bruce and the Formation of the Stock Branch

In 1864 Augustus Morris, a convert to curing rather than killing, steered yet another Scab Bill through the New South Wales Parliament.[32] The eventual act set the scene for the eradication of scab in the colony and laid the foundations of livestock-disease control, eventually in all the Australian colonies, for the rest of the century.

The 1864 act provided for compulsory dipping rather than slaughter. This meant that effective containment could be maintained at a reasonably low cost. In fact, the new levy was set at half the rate of the old (and was only exacted on owners of five hundred sheep and more), yielding nine thousand pounds in 1864. Of equal significance, the other provisions of the act represented a nice mix of the local control desired by pastoralists and the central power necessary to coordinate colonywide public action. A permanent inspectorate was established, the regional members to be appointed by district scab boards made up of the elected representatives of shipowners. These scab directors, as they were termed, could act as and had supervisory powers over the local inspectors. However, the latter were paid from the central Scab Fund and came under the direction of a government-appointed chief inspector of sheep, who was directly responsible for the coastal region and thus the markets that had been the chief source of infestation.[33]

There was obvious potential for dissention in this shared control. This was soon demonstrated in a number of conflicts between the chief inspector and local scab boards[34] that continued into the twentieth century.[35]

However, such divisions proved incidental in the short and longer run, in large part due to the personality of Alexander Bruce, who became the first chief inspector of sheep in 1864. In fact, it was largely owing to Bruce that, after the successful eradication of scab, the machinery established under the act evolved further into a broad system of disease control—and more.

Bruce was a Scottish migrant who had already failed twice as a pastoralist. This was no reflection of his abilities; he was intelligent, of broad interests, and a workaholic. He also had connections, including family, with pastoralists in the Riverina, and these assisted his path into public appointments, first as a pound keeper in the Riverina in 1858, and then as an inspector of cattle on the Murray in 1861.[36] In this role, he led a campaign to prevent the introduction of contagious bovine pleuropneumonia (CBPP) from Victoria across the Murray into New South Wales. The campaign failed, despite the slaughter of thousands of cattle at considerable cost to the government, but Bruce retained his inspectorship, continuing to search for answers to the disease.[37] With pastoralists now fully aware of the dangers of importing disease, his combination of inspectorships of sheep and cattle anticipated his appointment as chief inspector of stock and the formal inception of the stock branch two years later.

Bruce tackled the scab outbreak with what was to prove typical drive and energy. His first report as chief inspector demonstrated the efficacy of the new mixture: some three hundred fifty thousand infested sheep had been dealt with in the past two years; forty thousand had been destroyed, the rest being successfully dipped.[38] By the end of 1865, Bruce could claim that "the sheep in this Colony are entirely free from scab." He was wrong, but the two outbreaks of 1866 reinforced his message on the need for continuing vigilance.[39] More importantly, the other colonies now had a control model they could copy, thus reducing the risk of reinfestation in the longer term.[40] Finally, the mechanisms established in New South Wales to eradicate scab continued to evolve to meet pastoralist objectives in a broader perspective.

Eradication gave Bruce and his regulatory apparatus credibility and prestige. In 1865, the stock branch was still overtly intended to meet the one disease. It quickly, however, found new functions, and its development afterward matched the piecemeal nature of its emergence. With pastoralists now fully aware of the costs of imported disease, it became responsible for monitoring all livestock imports. The Brands Act of 1866 made it responsible for allocating cattle brands while Livestock Diseases Acts in 1876 and 1878 gave the inspectorate and local boards powers over traveling stock and reserves.[41] Such functions were consolidated in the Pastures and

Stock Protection Act of 1880, which embodied the same division of powers between the branch and local boards elected by landowners established under the Scab Acts.[42] The Scab Fund remained the branch's chief source of funds, even though it was no longer used to meet scab, while other income came from a variety of charges, licenses, and fines. Insofar as there was any coherence to the structure of the branch, then it lay in the position of Alexander Bruce. However, while the Scab Fund gave him autonomy within the bureaucratic structure of colonial government, Bruce was ever aware of who his masters were.

The prime objective of the branch was to meet threats to the profitability of pastoralism. Its focus was largely but never entirely veterinary; a broader theme was meeting the consequences of a dynamic process of ecological imperialism. Its greatest success was in preventing further livestock-disease imports, while it played a coordinating role in the control of those already established and also became a key mechanism in vermin destruction, especially, at least initially, in meeting the spread of the rabbit plague from the 1870s. Bruce, as chief inspector, usually tried to set the agenda for concerted action, using his annual reports to warn of looming external disease threats, of the internal prevalence of CBPP and anthrax, and of the dangers of "over-stocking" the land. However, he was always highly sensitive to pastoralist opinion, maintaining a tradition of extensive consultation on policy. Thus, as an enthusiast for what was known as "tail inoculation"[43] against CBPP, Bruce wanted to establish a compulsory program under the auspices of the branch and conducted a number of surveys seeking pastoralist support. The latter, however, although convinced of the utility of inoculation as a prophylactic, were skeptical (probably rightly) of the benefits of a full-scale program, and this never happened.[44]

There were thus limits to what Bruce could achieve, while he always had limited and often inadequate resources to deploy. Further, the nature of the stock branch became increasingly at odds with the growing trend toward the utilization of scientific and professional expertise in meeting the disease and other ecological threats that continued to grow in importance.

Decline of the Stock Branch

Resource problems first eroded the original model of the stock branch. Income from the levy peaked in 1891; thereafter, it declined with stock numbers, to fall by 30 percent during a deep depression for Australian pastoralism. In the 1890s stockowners faced not only a continuing rabbit plague, lower wool prices, labor problems, and a sequence of droughts but also an increased incidence of existing disease and the arrival of new ones (notably fly-strike in the 1890s). This laid bare a fundamental problem: the ability of the branch to mount its services declined at the very time when these were most needed. The branch's income had to be supplemented from general revenue in 1898, despite reductions in the size of and payments to the inspectorate.[45]

This was at the same time that inspectors were first required to demonstrate some professional competence. In 1888, new regulations made provision for entry by examination, although this was not actually enforced until 1895. The first successful examinee was a graduate of the Royal Veterinary College in London, a testimony to the high veterinary content of the examination and a portent of twentieth-century trends.[46] During his regime, Bruce had been an enthusiast for veterinary science, but less so for its practitioners, unsurprisingly, given his experience of their competence.[47] His appointment of J. D. Stewart as veterinary adviser in 1898, however, was a further landmark in the shift to professional scientific expertise. Stewart, after an active career that included appointment as the first professor of veterinary science at Sydney University, was to become chief inspector within the decade.

Pastoralist control over the system was also eroded in this period. The chief overt sign was the removal of their role, as directors of local stock boards, in the appointment of stock inspectors in 1902. A full veterinary qualification was required for the position by the 1920s; the name itself was changed from "stock" to "veterinary inspector" in 1937. They became officers in the 1960s.[48] Use of the term "stock branch" had ceased by the 1930s, and the Veterinary Service of the New South Wales Department of Agriculture had come close to resembling the European model.

THE ESTABLISHMENT of the stock branch in New South Wales was just one of a stream of innovations that established the global ascendancy of Australian pastoralism in the nineteenth century.[49] Its origins lay in the capacity of a small group of highly motivated capitalists to act creatively in the face of ecological challenges to the profitability of livestock. Their ability and willingness to pay for the measures necessary to control livestock disease was also unusual. In the early twenty-first century, the concept of farmers paying for publicly mounted services has become more common, but they seldom have the degree of control over such services enjoyed by nineteenth-century Australian pastoralists.

The ultimately fatal weakness of the stock branch was financial. The ability of stockowners to finance the disease control the branch provided declined in the conditions when such controls were most needed. Nevertheless, it has a number of contemporary legacies. One is the continuing existence of the rural lands protection boards in New South Wales, direct descendents of the scab boards, which still levy land- and stockowners to provide a variety of local services. Another lies in the continuing tradition of stockowner participation in disease and feral-animal control. This does not guarantee success, as contemporary problems with ovine Johnes disease in New South Wales demonstrate,[50] but generally contributes to effectiveness. The final legacy is embodied in the strictness of quarantine regulations that have helped to maintain the relative freedom of Australia from disease and continues to underpin its ascendancy in the global market for livestock and livestock produce.

Notes

1. Sylvie Lepage, "Too Many Sheep, Too Little Profit," *Le Monde*, 8 June 1993; translated in the *Guardian Weekly*, 27 June 1993.

2. See, for example, John Fisher, "To Kill or Not to Kill: The Eradication of Contagious Bovine Pleuro-pneumonia in Western Europe," *Medical History* 47, no. 3 (2003): 314–31; W. Schönherr, "History of Veterinary Public Health in Europe in the 19th Century," *Revue scientifique et technique* 10 (1991): 985–94; Daniel Gilfoyle, "Veterinary Science and Public Policy at the Cape Colony, 1877–1910" (PhD diss., University of Oxford, 2002); Ole H. V. Stalheim, *The Winning of Animal Health: 100 Years of Veterinary Medicine* (Ames: Iowa State University Press, 1994); T. Dukes and N. McAnninch, "Health of Animals Branch, Agriculture Canada: A Look at the Past," *Canadian Veterinary Journal* 33, no. 1 (1992): 58–64.

3. The most recent biography is Michael Duffy, *Man of Honour: John Macarthur* (Sydney: Macmillan, 2003).

4. Alfred Crosby, *Ecological Imperialism: The Biological Expansion of Europe, 900–1900* (Cambridge: Cambridge University Press, 1986).

5. See John Cheyne Garran and Les White, *Merinos, Myths and Macarthurs* (Canberra: Australian National University Press, 1985); Charles Massy, *The Australian Merino* (Melbourne: Viking O'Neil, 1990).

6. Stephen Henry Roberts, *The Squatting Age in Australia, 1835–1847* (Melbourne: Melbourne University Press, 1964; facsimile reprint of 1932 edition); John C. Weaver, "Beyond the Fatal Shore: Pastoral Squatting and the Occupation of Australia, 1828 to 1842," *American Historical Review* 101, no. 4 (1996): 986–1004.

7. Garran and White, *Merinos, Myths*, 206–10.

8. *Statistical Register of New South Wales 1859* (Sydney: Government Printer, 1860).

9. See, for example, the debates over Keith Windschuttle's *The Fabrication of Aboriginal History* (Sydney: Macleay Press, 2002), in the Australian Council of Professional Historians Association's Inc. discussion forum at http://www. historians.org.au/discus/.

10. Robert Dawson, *The Present State of Australia* (London: Smith Elder, 1830), 417.

11. The account below is based largely on John Fisher, "Technical and Institutional Innovation in Nineteenth-Century Australian Pastoralism: The Eradication of *Psoroptic Mange* in Australia," *Journal of the Royal Australian Historical Society* 84, no. 2 (1998): 38–55.

12. Walter James Malden, *Sheep Raising and Shepherding* (London: MacDonald and Martin, 1899), iii, 201.

13. Fisher, "Technical and Institutional Innovation," 41–44.

14. Bourke to Goderich, 30 October 1832, in *Historical Records of Australia* (Sydney: Library Committee of the Commonwealth Parliament XVI, 1925), 782–83.

15. See Peter Mylrea, "Catarrh in Sheep," *Australian Veterinary Journal* 69, no. 9 (1992): 298–300.

16. The Settled Districts, also known as the Nineteen Counties, were the areas of earliest white occupation. The major regions were the Sydney Basin and the Hunter Valley.

17. "Report of the Select Committee on Scab and Catarrh in Sheep," *Sydney Morning Herald,* 6 October 1854.

18. Legislative Council Debates in *Sydney Morning Herald,* 6 and 14 October, and 18 November 1854.

19. Return on Scab-in-Sheep Act of 1854, *NSW Legislative Council Votes & Proceedings* (*NSWLC V&P*), vol. 2 (Sydney: Government Printer, 1855); *Sydney Morning Herald,* 1 August 1855.

20. Select Committee on Scab in Sheep, 1858, Evidence 34–35, in *NSWLA V&P*, vol. 2 (Sydney: Government Printer, 1858); see also John Fisher, "Meeting Sheep Scab on an Early Victorian Pastoral Property: A Story from the *Clyde Company Papers*," *Australian Veterinary History Record* 24 (1999): 4–11; Philip Laurence Brown, ed., *Clyde Company Papers*, vol. 3 (London: Oxford University Press, 1958), 523.

21. The point is best set out in Graham J. Abbott, *The Pastoral Age* (Melbourne: Macmillan, 1971).

22. See, for example, *Sydney Morning Herald,* 11 November 1854.

23. Report of the Select Committee on Scab and Catarrh in Sheep, 6, in *NSWLA V&P 1857*, vol. 2 (Sydney: Government Printer, 1858).

24. Select Committee on Scab in Sheep, appendix A, in *NSWLC V&P 1858*, vol. 3 (Sydney: Government Printer, 1858–59), 1405.

25. Report of the Select Committee on Scab and Catarrh in Sheep, 6.

26. Select Committee on Scab in Sheep, 1858, Evidence, 20–23.

27. *Sydney Morning Herald,* 20 December 1861 and 11 February 1862 (extract from the Riverina *Pastoral Times*).

28. *Sydney Morning Herald,* 9 July 1863.

29. See James Charles Hamilton, *Pioneering Days in Western Victoria: A Narrative of Early Station Life* (Melbourne: Macmillan, 1923), 28–29; Margaret Kiddle, *Men of Yesterday* (Melbourne: Melbourne University Press, 1961), 63–64.

30. Brown, *Clyde Company Papers,* vol. 4, 326, 382.

31. Noel G. Butlin, *Investment in Australian Economic Development, 1861–1900* (Cambridge: Cambridge University Press, 1964), 72–76; Geoff Raby, *Making Rural Australia: An Economic History of Technical and Institutional Creativity, 1788–1860* (Melbourne: Oxford University Press, 1996), 100–11.

32. *Sydney Morning Herald,* 27 and 28 November 1863, 10 and 11 December 1863, and 8 and 18 January 1864. His success proved too late to save him from bankruptcy; see entry on Augustus Morris (1820?–1895) in *Australian Dictionary of Biography, 1851–1900,* vol. 5 (Melbourne: Melbourne University Press, 1966), 292.

33. *Sydney Morning Herald,* 14 June 1864; see also William Lloyd Hindmarsh, "Historical Records of the Veterinary Profession in Australia, Part 2," *Australian Veterinary Journal* 47, no. 12 (1971): 510–11.

34. For an example, see Scab Inspector, Warialda, Correspondence, in *NSWLA V&P 1865,* vol. 2 (Sydney: Government Printer, 1865). The Warialda directors saw no necessity to appoint an additional inspector at the Queensland border—at their own expense.

35. E. A. Lucas, "Our Institute, A Short History," *Institute of Inspectors of Stock of NSW Year Book* (1945): 7–16.

36. Edward John McBarron and Herbert Robert Seddon, "Aspects of the Life of Alexander Bruce, First Chief Inspector of Stock, New South Wales (1864–1902)," *Veterinary Inspector* 30 (1966): 77–83.

37. Edward John McBarron, "Alexander Bruce, Inspector of Cattle, and the Advent of *Pleuro-Pneumonia Contagiosa* to New South Wales," *Australian Veterinary Journal* 29, no. 4 (1952): 99–105.

38. Report from the chief inspector of sheep, 1865, *NSWLA V&P 1865,* 2:1–6.

39. Reports from chief inspector of sheep, *NSWLA V&P 1865,* 5 (1866), and *NSWLA V&P 1867,* 4 (1867–68).

40. For developments in other colonies, see E. M. Pullar, "Sheep Scab in Victoria, 1834–76," *Victorian Veterinary Proceedings* 26 (1967–68): 11–16; W. R. Smith, "History of the Control of Some Diseases of Sheep and Cattle in South

Australia," *Australian Veterinary Journal* 60, no. 4 (1983): 122–23; Leslie G. Newton, "Historical and Administrative Aspects of Pleuro-pneumonia's Eradication in Queensland," *Public Administration* 33 (1974): 43–59.

41. Hindmarsh, "Historical Records," 511–13.

42. John Fisher, "Pastoral Development and the Veterinary Profession in Australia, 1850–1900," *Australian Veterinary Journal* 72, no. 8 (1995): 126–31.

43. Cattle were "inoculated" in the tail with serous material from a diseased animal in a manner that proved an effective prophylactic. See C. Huygelen, "Louis Willems (1822–1907) and the Immunization against Contagious Bovine Pleuropneumonia: An Evaluation," *Verhandelingen van de Koninklijke Academie voor Geneeskunde van België* 59 (1997): 237–52.

44. Bruce participated in an Australian process of product innovation that built on the original work of Louis Willems; see Leslie G. Newton and Ronald Norris, *Clearing a Continent: The Eradication of Contagious Bovine Pleuropneumonia from Australia* (Melbourne: CSIRO, 2000); John Fisher, "*Every Man His Own Farrier* in Australia: The Origins and Growth of a Veterinary Business in Colonial New South Wales," *Argos* 23 (2000): 140–42.

45. Fisher, "Pastoral Development," 129.

46. William Lloyd Hindmarsh, "The Evolution of the Stock Inspector Service in the Nineteenth Century," *Veterinary Inspector* 31 (1967): 78.

47. See John Fisher, "Foot and Mouth Disease in Australia," *Australian Veterinary Journal* 61, no. 4 (May 1984): 158–61; and idem, "The Origins of Animal Quarantine in Australia," *Australian Veterinary Journal* 78, no. 7 (June 2000): 478–82.

48. Fisher, "Pastoral Development," 129–30; Lucas, "Our Institute," 7–16; Hindmarsh, "Stock Inspector Service," 73–79; E. A. Farleigh, "The Pastures Protection Board," *Institute of Inspectors of Stock of N.S.W. Year Book* (1937): 94–96; J. G. Johnston, "Disease Control in N.S.W.," *Institute of Inspectors of Stock of N.S.W. Year Book* (1953): 7–10; Peter Mylrea, *In the Service of Agriculture: A Centennial History of the New South Wales Department of Agriculture* (Sydney: NSW Agriculture and Fisheries, 1990).

49. For others, see Gerald Walsh, *Pioneering Days: People and Innovations in Australia's Rural Past* (Sydney: Allen and Unwin, 1993); and Raby, *Making Rural Australia*.

50. See New South Wales Department of Primary Industries, "Ovine Johne's Disease," *Agriculture*, http://www.agric.nsw.gov.au/reader/ojd.

Chapter 11

Holding Water in Bamboo Buckets

Agricultural Science, Livestock Breeding, and Veterinary Medicine in Colonial Manchuria

Robert John Perrins

A BITTER Siberian wind greeted delegates as they arrived at the army supply depot on the outskirts of the Manchurian port city of Dairen just after noon on 26 January 1914. Once inside the compound's main office building, the twenty scientists and administrators were ushered into a makeshift meeting room where the station's military commander welcomed them to what was to be the inaugural meeting of the South Manchuria Veterinary Association (*Nan Manshū jūi kyōkai*).[1] Over the course of the half-day conference, the representatives from the various branches of the Japanese colonial administration listened to more than a half-dozen presentations of scientific and public health papers before concluding their meeting with a discussion of the need to coordinate their efforts on a number of fronts. To this end, it was decided that several branches of the colonial system, including the Kwantung Government-General (*Kantō-tōtokufu*);[2] the regional garrison of the Kwantung Army (*Kantō-gun*);[3] the South Manchuria Railway Company (*Mantetsu*);[4] and municipal police forces in the cities of Dairen, Ryōjun (Port Arthur), Mukden, Andong, and Jinzhou would cooperate on a number of projects, including the prevention and monitoring of animal diseases in the region, as well as sharing research on both animal diseases and the development of improved breeds of livestock.[5]

The veterinarians, biologists, police officers, physicians, and railway managers who were in attendance at this meeting were keenly aware of the pressing need for such action, having just heard a lengthy report on the outbreak of cattle plague, or rinderpest, that had been presented by the veterinary surgeon attached to the Japanese consulate in the city of Andong on the Manchurian–Korean border.[6] Both the content and tone of this report struck a nerve with the colonial officials attending the conference. The audience was told how rinderpest had first appeared in eastern Jilin Province and how it had moved steadily toward the Manchurian–Korean border between the spring of 1912 and the summer of 1913. The author of the paper confidently, but erroneously, claimed that the disease had been transported down the Hun and Yalu rivers by Chinese boatmen who had purchased meat of diseased cattle in the remote interior and who had then somehow transmitted the illness to healthy animals during their travels. Almost twenty thousand cattle had died during the outbreak, and there were fears that the situation would worsen as the weather began to warm in the spring and the herds of local cattle began to intermix as they moved from their winter shelters to new grazing lands.

While the disease had been stamped out on the Korean side of the Yalu, due to the combination of a vigorous rural surveillance system and a mandatory vaccination program for all cattle in the new Japanese colony, the situation in Manchuria, the audience was warned, was not as promising. The reason given for the continued threat of the disease was the "callousness of the Chinese" authorities and farmers who were resisting both the inspection and vaccination of their livestock.[7] The health of the region's livestock and the local rural economy were threatened with ruin, the veterinarian from Andong argued, unless the colonial authorities began to coordinate their efforts in the battle against cattle plague.[8] Reflecting on this epizootic crisis just over a decade later, Dr. Kasai Katsuhiro, one of the region's leading veterinarians and the founding superintendent of the South Manchuria Railway Company's Cattle Disease Institute in Mukden (Hōten jūeki kenkyūjo), lamented that any efforts to improve the rural economy or develop better livestock breeds during this period had been almost useless and akin to "trying to hold water in a bamboo bucket."[9] Kasai's pessimistic assessment of this period was due to the prevalence of not only livestock diseases such as rinderpest, glanders, bovine pleuropneumonia[10] and sheep pox[11] in southern Manchuria but also to what he believed to be the ignorance, fear, and resistance to change on the part of most of the region's Chinese population.

The Japanese administrators and scientists who attended the conference on that cold January afternoon in 1914, however, were not resistant to change or suspicious of "modern" science. In fact, by the time the inaugural meeting of the colonial veterinary association concluded, all of the attending delegates were convinced that decisive action needed to be taken. Confident that they were armed with the best "modern" scientific knowledge and expertise, these agents of Japanese colonialism in Manchuria envisioned that they were heroically manning the barricades against invading armies of microbes that threatened both the economy and health of the fledgling colony.[12] Over the course of the first four decades of the twentieth century, Japanese veterinarians, agronomists, livestock breeders, and colonial administrators attempted to demonstrate Japan's leadership role in connecting Manchuria, China, and the rest of East Asia to the wonders of "modernity."[13] The development of veterinary medicine, new agricultural crops, and an expanded livestock economy played important supporting roles in the history of Japan's Manchurian empire and were closely tied to the broader policies of "scientific colonialism" and the development of a Western-styled system of public health.

This sense of confidence on the borders of the empire was in many ways a reflection of Japan's national achievements during the Meiji era (1868–1912). During this period, Japanese society, politics, and industry had been transformed and modernized under the slogan of *bunmei-kaika* ("civilization and enlightenment"). The result was the creation of a strong nation, as well as a new nationalism that was, in part, related to extending Japan's presence abroad through the acquisition of colonies.[14] Although a latecomer to the colonial game, by the early 1900s Japan controlled a number of overseas territories, including Taiwan, the Kwantung leasehold in southern Manchuria, and Korea. Within the space of a single generation, Japan had emerged as a growing imperial power in East Asia. The prize for many Japanese imperial dreamers and planners in the early twentieth century would eventually be the puppet state of Manzhouguo that was created by the Japanese military in 1932.[15]

Colonial Manchuria

Manchuria was the great Asian frontier and one of the most contested borderlands in the world during the late nineteenth and early twentieth centuries.[16] As a geographical space, Manchuria was in many ways a constantly shifting creation of foreign imperialists. In fact, the Chinese historically referred to the temperate region of steppes, farmland, and forests not as Manchuria but as the "three eastern provinces" (*Dongsansheng*) and since 1945, have simply called the territory the "northeast" (*Dongbei*).[17] For foreign powers, however, during the 1890s and early 1900s, Manchuria was conceived as being comprised of the Qing, and later Republican, provinces of Heilongjiang, Jilin, and Fengtian. After 1932 this "imagined place" was expanded by Japanese imperialists to include the Chinese province of Rehe (Jehol) and parts of eastern Inner Mongolia referred to as Xing'an.[18] Lastly, in the mid-1930s, the Manzhouguo regime reorganized their nation's political map, dividing Manchuria's original three provinces into smaller units along with the newly added territories to the west, for a grand total of nineteen provinces with a total area of more than 1.3 million square kilometers—making Manzhouguo geographically larger than Japan and all of its formal colonial empire combined.[19]

Regardless of its amorphous boundaries between the late Meiji and Showa eras, Manchuria was consistently viewed in Japan as a strategic buffer to Russian expansion in northeast Asia, a source of raw materials, and a potential home to millions of Japanese settlers. Manchuria, or *Manshū*, grew to have a psychological hold over much of the Japanese population during the 1930s and early 1940s as the colony was portrayed as both a modern utopia and a vital storehouse supplying natural resources, industrial products, and agricultural goods to the imperial body.[20]

Early Japanese Efforts to Develop Southern Manchuria

One of the most important figures in the formulation of the vision for the development of the early Japanese Empire was Gotō Shinpei, a physician by training and the first chief civilian administrator (*Minsei Chōkan*) in colonial Taiwan.[21] During his decade of service in Taiwan (1895–1905), Gotō sought to develop the island as a modern colony using both rational scientific principles and the "biological principles" of imperialism—the idea that the local economy and society were to be developed, or "uplifted," by the Japanese authorities.[22] After serving in Taiwan, Gotō was presented with an opportunity to pursue his vision of colonial development in Japan's newest colony when he was appointed the first president of the South Manchuria Railway Company (SMR) in 1906.

As president of the SMR, Gotō wielded a great deal of power in southern Manchuria, and while the Kwantung Army was charged with protecting the leasehold and its railway network, he was given almost free rein to develop the colony's economy and social policies.[23] While the majority of the SMR's efforts were clearly focused on developing the region's transportation infrastructure and industrial base, Japan's largest corporation in southern Manchuria was also interested in the rural sector.[24] Following

Gotō's vision of rational colonialism, the Mantetsu company initiated dozens of scientific studies of Manchuria's soil, climate, crops, and livestock during the first few years of its existence.[25] This early work was focused on investigating "traditional" Chinese farming in the colony with the hope of developing strategies to increase production of cereals and livestock that would then be shipped as freight by the Mantetsu railway.[26]

These early agricultural studies, however, were limited in scope in that they focused on the farms on the Liaodong Peninsula in southern Manchuria and on crop diseases, paying little attention to animal issues. This soon changed in the fall of 1910 when reports began to emerge of cases of plague breaking out in the northern region bordering Siberia and Mongolia.[27] Instead of being described as a potential agricultural Eden, Manchuria now became a land to be feared—literally a "plague-land" where natural reservoirs of the disease lurked among the local animal populations.[28]

Between November 1910 and April 1911, pneumonic plague swept through the crowded cities along Manchuria's railways killing more than sixty thousand persons.[29] Among the Japanese agencies that worked to combat the plague were the Kwantung Army and local consular police, as well as dozens of scientists and sanitary officers employed by the SMR. Not only did the local Chinese and Japanese authorities work to quarantine the population of sojourning laborers from the provinces south of the Great Wall, but they also conducted field research into the life cycle of the native plague carriers, the tarbagans, or Mongolian marmots (*Marmota sibirica*)—rodents that are similar to North American prairie dogs—that lived on the steppes of northwestern Manchuria.[30]

While the originators of Japan's colonial veterinary service in Manchuria were the scientists who conducted these early studies of plague and its rodent carriers, during the late 1910s Japanese biologists, bacteriologists, and veterinarians expanded their work to include studies of animal diseases such as rinderpest, glanders,[31] and anthrax,[32] diseases that were endemic to the region.[33] In the years immediately following the Manchurian plague epidemic, Japanese veterinarians dedicated themselves both to studying diseases of wild and domesticated animals and to developing new breeds of livestock with which they planned to revamp the colony's agricultural economy.

Prior to the founding of Manzhouguo in 1932, the Japanese colonial presence in Manchuria was centered in the Kwantung leasehold in southern Manchuria and along the SMR. Following the vision of Gotō Shinpei, the Mantetsu company, together with the Kwantung administration, pursued a policy of "scientific colonialism" that emphasized the development of

Manchuria using science, technology, and engineering. As part of their shared strategy, the SMR and the Kwantung administration established dozens of hospitals, research institutes, and experimental farms in an effort to combat diseases (of both humans and animals), as well as to engineer improvements to the fledgling colony's agricultural economy.[34] During the first two decades of Japanese rule in southern Manchuria, the Kwantung administration established three agricultural research stations, and the SMR founded a total of nine scientific institutes, including two dedicated to agricultural and livestock research, as well as a central laboratory, a hygiene institute, and its main research office in Dairen.[35]

Experimental Farms

The first experimental farm in Manchuria was established by the Kwantung administration in 1907 on a small plot of land in the center of the port city of Dairen.[36] Agronomists and veterinarians were recruited from Japan to work at this facility, where they spent a decade investigating methods to improve both the region's fruit production and livestock.[37] As Dairen grew into the largest urban center in the colony, this experimental farm was forced to relocate to the main Mantetsu compound in the Shakakō district in the western end of the city. Here the Japanese staff continued their research, as well as delivering free public lectures to local Chinese and Japanese farmers.[38] In 1924, this experimental farm was relocated for the last time to an eighty-four-hectare site near the town of Jinzhou, and four years later the station's administration was incorporated into the general Agricultural Experimental Farm under the joint control of the Kwantung administration and the SMR.[39] At the farm in Jinzhou, Japanese scientists conducted research during the 1910s and 1920s on cattle, sheep, and swine breeding, sericulture, and the improvement of vegetable and fruit crops.[40] It is important to note that this facility was not only one of the first agricultural institutions established in the empire but also that it predated the founding of similar facilities in Japan.[41]

The largest agricultural station, encompassing a compound of more than two hundred hectares, was founded in 1913 by the SMR near the town of Gongzhuling, 650 kilometers north of the port of Dairen at a cost of more than half a million yen.[42] This institute had a staff of almost one hundred, including research botanists and agronomists; dozens of laborers who tended the experimental fields and orchards, as well as the swine, cattle, sheep, and horses housed in the institute's barns; several laboratory assistants and chemists; and over a dozen veterinarians.[43] The goal of the Gongzhuling farm was to improve what the Japanese held to be paltry yields of local crops, as well as the puny native livestock raised by the region's Chinese farmers. To this end, the station developed several strains of soybeans and fruit trees, along with several "improved" breeds of sheep.[44] It had become clear to some SMR managers that if the Manchurian colony was to thrive, its rural sector had to be modernized and developed using all the tools of modern science. Such efforts, however, required a "civilizing vision" of improving the colony, as well as financial support.

By the early 1930s, the SMR had already spent more than one million yen on the Gongzhuling facility, and its annual budget had grown to in excess of 350,000 yen. However, it should be noted that this figure amounted to less than 0.2 percent of the annual Mantetsu budget. Despite glowing descriptions of Manchuria as the "Garden of China" and related propaganda claims that Japan was desperately attempting to modernize the region's farming techniques and domestic breeds of livestock, the colonizers' direct investment in the agriculture sector should not be exaggerated.[45] Using the figures from the 1930–31 Mantetsu budget, for example, one notes that expenditures for maintaining the region's railway network amounted to 36.77 million yen (22.1 percent of total expenditures); the coal mines at Fushun cost 60.63 million yen (36.4 percent of total) to operate; the iron works at Anshan claimed 7.31 million yen (4.4 percent of total); and the development of the region's main commercial port at Dairen cost the SMR 6.74 million yen (4 percent of total) a year.[46] Improving Manchuria's livestock, fruit, and cereals may have been the primary focus of the veterinarians and crop scientists at Gongzhuling, but such efforts paled in comparison to their employer's strategy that focused on industrial development.[47]

During the first decade of their operations, the various experimental farms that dotted southern Manchuria concentrated their efforts on the development not of new breeds of cattle, sheep, or swine but rather of improved varieties of fruit trees and soybeans. To assist in feeding the growing urban population in the cities and towns along the SMR, the Kwantung administration instructed the stations' staffs to develop new varieties of apple, peach, pear, and cherry trees, along with a diversified staple of vegetable crops.[48] Between 1910 and the early 1920s, a virtual forest of 590,000 fruit-tree seedlings, utilizing new stock imported from the United States, were developed in the orchards at the Japanese-run agricultural stations and distributed free of charge to the region's cultivators.[49] This effort to enhance the colony's fruit crops, however, was dwarfed by the farms' program to improve Manchuria's soybean crop.

Japanese crop scientists and biologists targeted the lowly soybean for colonial improvement, and this "heroic" work was outlined in the SMR's report of 1929:

Among many experiments, most serious attention has been paid to improving the Manchuria bean. After a series of experiments, this station [Gongzhuling] succeeded in obtaining, by means of selection, four superior kinds of beans. Experiment further continued, and it has finally advanced to two best kinds which have been adopted as the standard of the improved beans. . . . In 1922, a nursery farm for the improved seed beans, about 174 acres [70 hectares] in area, was laid out at Changchun and Gaiyuan, with a view to supplying the improved seed to a more extended area further from the Railway Zone, eventually to cover the whole of Manchuria.[50]

These engineered beans produced 10 to 20 percent more crop per plant, and 8 percent more bean oil than the unaltered native stock.[51] The soybean varieties that were developed on the SMR's experimental farms spread swiftly across Manchuria during the late 1910s and early 1920s and were, in part, responsible for the huge bumper crops that supplied much of North America and western Europe with soybean oil, bean-cake fertilizer, and soy protein during World War I.[52]

Colonial Livestock Breeding Programs

Following the success of their early work cultivating improved varieties of fruit trees, cereals, and beans, the SMR experimental farms shifted the focus of their activities in the early 1920s to the development of Manchuria's livestock. In an effort to improve the "slight builds and poor physiques" of the native breeds of domesticated animals, Japanese scientists and veterinarians initiated a number of ambitious crossbreeding programs.[53] While admitting that the Chinese farmers in the region had managed to raise sheep, cattle, and horses in the rugged environment of the Manchurian steppes for the past two hundred years, a Japanese observer noted, "The methods used for the breeding and handling of livestock in Manchuria are primitive. . . . Therefore, although these domestic animals are sturdy, their quality is course, and there is much need for improvement."[54] By 1921, the main experimental station at Gongzhuling, along with the branch farm located near the town of Xiongyaozheng, began to engineer new hybrids of cattle, horses, mules, sheep, and swine, with the intent of improving the local economy as well as directly assisting local Chinese farmers by distributing the superior breeding animals at very low costs.[55]

Thousands of breeding animals were imported from throughout the Japanese Empire and even further afield for this purpose. Korean cattle, which had higher fat content in their meat, were crossbred with native Chinese cows. Hackneys and Anglo-Arabian breeds of horses were imported from Japan and bred with shorter Manchurian horses. The famous large mules from across the Bohai Sea in China's Shandong Province were shipped by the hundreds to breeding stations along the SMR line. Berkshires from both the United States and Britain were imported in an effort to improve the local swine population. And lastly, more than five hundred Merino sheep from the United States (favored for the quality of their wool), as well as three hundred Corriedales from Australia (preferred for their meat) were purchased during the 1920s by Mantetsu as breeding stock for its experimental farms.[56]

The effort to improve the native Manchurian and Mongolian sheep stocks was further enhanced by the establishment of additional breeding stations at the towns of Heishandun and Da'erhan in the mid-1920s. After the founding of Manzhouguo in 1932, three more sheep farms were established by the General Directorate of the Manzhouguo State Railway Ministry at Baizhengzi, Yangzhuanzi, and Baijia in central Manchuria.[57] It was wildly estimated by the colonial authorities that, when the two million sheep in Manchuria and Inner Mongolia had been "improved" by Mantetsu's breeding program, the region's total annual wool production would increase from approximately twenty-four hundred metric tons to more than nine thousand metric tons.[58]

During the 1920s, the veterinary staff at the SMR experimental stations imported fresh stocks of Merinos, Southdowns, and Shropshires from Great Britain, the United States, and Australia to crossbreed with native animals. The result of the first crossbreeding of Merinos with native stock resulted in an increase in the output of wool of two-year-old sheep from 1.5 to 2.8 kilograms. The second breeding of Merinos with these hybrids resulted in a further increase to 4 kilograms. Improved wool yields were also obtained by crossing native sheep with Southdown sheep (from 1.5 to 2 kilograms) and with Shropshire sheep (from 1.5 to 2.7 kilograms).[59] While these efforts to engineer superior colonial sheep stocks were on the surface successful, one must be careful not to overstate their impact on the agricultural sector. Simply put, the numbers were pathetically small, as even by the late 1920s, less than one thousand hybrid sheep had successfully been bred, and as a result, there was simply not enough sheep available to assist local Chinese farmers who wished to improve their flocks.[60]

The showcase sheep-breeding program at Gongzhuling received a major boost after the creation of Manzhouguo in 1932, as the new puppet government sought to build upon the station's perceived earlier successes over the course of a new and ambitious thirty-year plan. This plan called for a total Manchurian herd of fifteen million head of sheep by 1967, all of which were to be of an improved Merino-cross variety.[61] However, by 1941, the last year for which statistics were published by the colonial government in Manchuria, only 3.7 million sheep existed, and less than 10 percent of these could in any way be classified as hybrid stocks.[62] While Japanese reports published during the 1940s blamed the plan's failure on their inability to import more foreign stock because of the Pacific War, the plan would have likely failed regardless of the military conflict.[63] The Merino-crossed sheep may have yielded more wool than did the native stock, but they were poorly adapted to life on the pastures of Manchuria and Mongolia. As the Australian sheep expert Ian Clunies Ross noted after a visit to the region in the mid-1930s, the hybrid sheep had difficulty surviving the harsh winters and required heated indoor pens as well as supplementary feed for up to six months of the year, making the increased wool yields simply uneconomical.[64]

Japanese authorities had taken great pride in claiming that the principal beneficiaries of their work developing Manchuria's agricultural potential during the 1910s and 1920s were the region's Chinese residents. Manchuria's Chinese farmer, they claimed, now grew more peaches, apples, and soybeans, and now raised larger and healthier animals than ever before. Admittedly, Japan did benefit indirectly since the colonial policy was "to develop Manchuria to the point where it can furnish her [Japan] with the raw materials which she [Japan] needs, which will naturally be accompanied by the development of an increased population with rapidly ascending buying power, which in turn will create a market in which Japan can sell her manufactures."[65] This said, however, what was emphasized in official reports and propaganda publications from this period was the claim that the crops and livestock that were being developed and bred were to be utilized on Chinese farms, by Chinese farmers.[66] This policy underwent a serious reworking in the early 1930s after the establishment of Manzhouguo.

Livestock and Japanese Settlement Plans

During the pre-Manzhouguo era, there had been small-scale attempts to settle Japanese farmers in southern Manchuria by the Kwantung government and the Mantetsu giant. These efforts, however, had all failed, as less than one thousand farmers had immigrated to the region during the quarter century of Japanese rule, and fewer than two hundred had remained as permanent settlers.[67] As early as 1932, the Department of Colonial Affairs in Tokyo devised a plan of settling one hundred thousand farming households in Manchuria over a ten-year period. This ambitious program was ultimately scaled down, and the Japanese Diet approved plans to a few, small trial colonies between the autumn of 1932 and spring of 1935.[68] While the total number of settlers who arrived during this first wave was only in the thousands, colonial authorities in Manchuria and ambitious imperialists in Japan were quick to note that these brave pioneers had established a real Japanese presence in rural Manchuria. The plan's supporters also boasted that these Japanese farmers had laid the foundation of a modern livestock economy in the colony, as the five initial settlements owned a "grand" total of 3,000 horses, 935 cattle, 4,655 sheep, and 3,400 hogs—figures that were a far cry from the millions still to be found in the hands of the region's Chinese farmers.[69]

Despite the modest achievements of these trial emigration projects, imperial planners, in both Tokyo and Manzhouguo's new capital of Xinjing, continued to develop grander settlement schemes that eventually emerged in 1936 as the "Million Households to Manchuria" (*Manshū e hyakumanto*) program.[70] Several months earlier, in October 1935, the Colonial Ministry published a document that came to set the standards for Japanese settlement in Manchuria. In the "Proposed Standards for the Management of Collective Farm Immigrants in North Manchuria," the Japanese yeoman farmers (*jisakunō*) were said to require, in addition to their twenty *chō* (20.25 hectares) of land, a plow, assorted farm tools, a wagon, one cow, one horse, one sow, and ten ewes.[71] By the mid-1930s, Japanese colonial policy had clearly shifted from developing livestock for the region's established Chinese farmers to equipping teams of sturdy new Japanese pioneers with equally sturdy livestock.

In the 1930s, along with the dream of settling one million Japanese households across the Manchurian steppes, the authorities in Xinjing also embarked on the ambitious goal of planning the overall development of their new colony. Inspired by Soviet and fascist state-controlled economic planning and models and by a desire to reverse the economic downturn that accompanied Manchuria's "declaration of independence" when tens of thousands of Chinese returned south of the Great Wall taking their livestock with them, Manzhouguo's new governors proclaimed the "General Outline of the Economic Construction Program of Manzhouguo," on January 1, 1937.[72] Two years in development, the five-year plan called for a governmental expenditure of 2.5 billion yuan to fund a program that would strengthen the Japan–Manzhouguo economic bloc, while enabling Manzhouguo to develop the resources necessary for both its own national defense and consumption, as well as to supplement the resources of Japan. While claiming that the agricultural sector, including the livestock economy, held not just "an important place" in the overall plan but was, in fact, "the backbone of Manzhouguo's national economy," only five million yuan were allocated to its development, while the majority of the plan's funds, totaling several hundred million yuan, were to be spent on large industrial projects such as mining, heavy industries, and electric-generating plants.[73]

The following livestock targets were set under Manzhouguo's first five-year plan: increasing the region's sheep stocks from 3 to 4.2 million; and adding 400,000 horses, 765,000 head of cattle, and 260,000 swine to the state's existing herds.[74] The introduction of a state-planned economy in Manchuria was a disaster for the region's agricultural sector, primarily because the beginning of the plan coincided with the commencement of the Sino–Japanese War. In the summer of 1937, Japanese forces invaded China, initiating a war that soon became a quagmire for the Japanese military. Over the course of the next eight years, the Japanese army requisitioned large numbers of Manchurian horses, mules, and donkeys, as well as meat products with which to feed its soldiers. Agricultural quotas simply could not be met, as breeding stock was hitched to military wagons and disappeared down country roads or was literally consumed by the wartime economy. While colonial propaganda continued to paint a rosy picture during the late 1930s and early 1940s, claiming livestock production figures that met or even exceeded targets,[75] the reality was severe hardship in the livestock industry and in rural communities, as both the state and military began to squeeze as much as possible from the colony.[76] In order to cope with the growing economic crisis, the new military rulers in Xinjing reluctantly turned once again to the expertise of their civilian rivals, the Mantetsu scientists and researchers.

The SMR had overseen not only much of the industrial and infrastructural development of the Japanese colony in Manchuria prior to the creation of Manzhouguo but also much of Japan's systematic and scientific investigation of the region. From 1907, when its main research bureau was founded, through to the early 1940s, the SMR prepared thousands of reports dealing with the company's enterprises, as well as Manchuria's natural environment, economy, hygiene, customs, labor conditions, and agriculture. In response to the agricultural crisis in the colony, Mantetsu researchers during the late 1930s and early 1940s conducted over 150 studies on conditions in rural Manchuria. Teams of veterinarians, biological scientists, economists, and anthropologists visited both Chinese and Japanese farms and villages throughout the nineteen provinces that now constituted Manzhouguo and surveyed the implementation of Xinjing's development strategies as well as the realities of rural life.[77]

Life in rural Manchuria during this period was challenging, as farmers faced droughts and floods on a regular basis, in addition to harsh winters for which few Japanese settlers or their hybrid livestock were adequately prepared.[78] Basic survival was a challenge for most of the Japanese farmers in Manchuria, and there was little prospect that they would be able to raise enough livestock to meet the targets set by the colonial planners sitting behind their desks in Xinjing or Tokyo. Perhaps in a good year, the Manchurian livestock sector could support the colony's needs for wool, meat, and animal labor, but there was little chance that it could fulfill its intended role as the main larder for Japan and the Imperial Japanese Army. If the realities of the colony's climate were not enough of a challenge, the war years also saw substantial declines in the values of livestock,[79] as well as periodic, but severe, outbreaks of contagious animal diseases, such as rinderpest and anthrax, among the region's herds.[80] Unfortunately, or possibly fortunately, for Japanese colonial dreamers, animal diseases were nothing new in Manchuria.

Veterinary Science and Animal Disease Research

By the mid-1920s, Manchuria ranked third in the world after the United States and Australia in terms of numbers of livestock but had the more dubious distinction of ranking first in the number of cases of livestock diseases, with annual losses totaling tens of millions of yen. To combat this threat to the colony, the SMR annually allocated just under one million yen, supporting over a hundred veterinarians, scientists, and support staff who worked at its growing number of experimental farms and breeding stations, as well as at its main agricultural facility at Gongzhuling.[81] With the growing importance of livestock to the region's economy, agricultural science in Manchuria began to shift its attention from crop research to studying and combating animal disease. The SMR's Cattle Disease Research Institute (*Hōten jūeki kenkyūjo*), later renamed the Manchurian Veterinary Institute (*Manshū Jūigakuin*), was established in 1925 in the railway town of Mukden in central Manchuria with an initial investment of 250,000 yen.[82] Over the next two decades, this facility supported the work of dozens of veterinarians who were engaged in studying and preventing diseases such as rinderpest, anthrax, foot-and-mouth disease,[83] hog cholera,[84] rabies,[85] and sheep pox. The Mukden institute had two divisions. The first involved general operations such as veterinary training, livestock inspection, and the manufacture and distribution of vaccines and serums. The second coordinated the activities of the various research departments both at the institute and on Mantetsu's experimental farms, as well as holding public lectures and workshops in rural communities on the topics of preventing and treating livestock diseases.[86]

The main focus of the work conducted by the research division of the Mukden institute centered on efforts to control rinderpest, or "steppe murrain"—the very disease that had united the region's Japanese scientists, medical officers, and administrators at the inaugural meeting of the South Manchuria Veterinary Association in January 1914. During the 1920s and 1930s, rinderpest remained endemic in Inner Mongolia and in western and central Manchuria, killing tens of thousands of cattle in frequent outbreaks.[87] Despite earlier efforts to develop a vaccine and inoculate cattle by the sanitation bureau of the Chinese Eastern Railway headquartered in the northern city of Harbin, it was not until the SMR began to produce hundreds of thousands of units of its own vaccine in the late 1920s, coupled with a public education and mandatory inoculation program supervised by a mobile veterinary corps, that the rate of cattle plague began to decline.[88] Director Kasai's bamboo basket was continuing to leak, but Japanese veterinarians, public health officials, and agricultural scientists could claim that they were at least trying to plug the holes.

THE HISTORY of veterinary science and the livestock industry in colonial Manchuria is part of the larger histories of Meiji-Showa Japan and its empire. When viewed through these broader historical lenses, the history of the evolving livestock strategy in Manchuria can be seen to relate to three major themes. The first was the widely held beliefs in the wonders of "modernity" and Western science that were closely tied to Japan's own nation-building during the Meiji era and Gotō Shinpei's vision of development and "rational colonialism." The second theme was introduction of an unrealized, but nevertheless officially planned, policy of settling hundreds of thousands of Japanese farmers in Manchuria—a vision that resulted in a shift in colonial policy from developing livestock for the ruled Chinese majority to creating super-breeds for colonizing Japanese pioneers. The third, and final, theme was the introduction of a state-planned and wartime economy in the puppet state of Manzhouguo that resulted in another change in colonial strategy to one that was aimed directly at developing resources for the empire and military. The veil of propaganda that continued to talk about "mutual development" became increasingly transparent during the 1930s and early 1940s as the livestock industry and veterinary science were more fully incorporated into a purely exploitative model of colonial rule in Manchuria.

Notes

The author would like to acknowledge Associated Medical Services and specifically the Hannah History of Medicine's Grant-in-Aid program, which provided the financial support that made this research possible. A note of thanks is also extended to Gillian Gunther, Madeline Fowler, and Hayden Lindskog for their assistance with the research for this chapter.

1. On the early history of veterinary services and efforts to control animal diseases in colonial Manchuria, see Gu Mingyi, *Riben qinzhan Lüda sishinian shi* (Shenyang: Liaoning renmin chubanshe, 1991), 339–48; M. Matsuo, *The Development of Science and Culture in Manchuria: Japan's Contributions* (Dairen: Research Committee of Pacific Relations in the South Manchuria Railway Company, n.d.), 1–22; South Manchuria Railway Company, *Report on Progress in Manchuria* (hereafter cited as *ROP*) (Dairen: South Manchuria Railway Company, 1931), 127–28, 158–62, and 228–29.

2. On the histories of the various Japanese colonial administrations in southern Manchuria between 1905 and 1945, see *Kantō-kyoku shisei sanjūnen shi* (Tokyo: Toppan insatsu kabushi kaisha, 1936), 61–78; and Kwantung Government, *The Kwantung Government: Its Functions and Works* (hereafter cited as *KGFW*) (Dairen: Manchuria Daily News, 1929), 16–21.

3. On the role of the Kwantung Army garrison in southern Manchuria and along the South Manchuria Railway, see Alvin D. Coox, "The Kwantung Army Dimension," in *The Japanese Informal Empire in China, 1895–1937*, ed. Peter Duus, Ramon H. Myers, and Mark R. Peattie (Princeton, NJ: Princeton University Press, 1989), 395–98.

4. Mantetsu is the abbreviation of the Japanese name of the railway company, *Minami Manshū tetsudō kabushiki kaisha*. For an introduction to the history of this colonial enterprise, see Ramon H. Myers, "Japanese Imperialism in Manchuria: The South Manchuria Railway Company, 1906–1933," in Duus, Myers, and Peattie, *Japanese Informal Empire*, 101–32.

5. *Manchuria Daily News* (hereafter cited as *MDN*), 7 January 1914.

6. Rinderpest, or cattle plague, is a disease caused by a virus from the family Paramyxoviridae and genus *Morbillivirus*.

7. *MDN*, 30 January, 1914.

8. On the perceived (and continuing) threat of rinderpest to the economy along the South Manchuria Railway and particularly in the Kwantung Leasehold on the Liaodong Peninsula where the port of Dairen was located, see *MDN*, 19 January 1914; the series of articles published by Dr. Kasai Katsuhiro titled, "Cattle Epidemics in Manchuria and Mongolia and How to Prevent Them," which were published in serial form in the *MDN* between 1 and 11

January 1926; and Inoue Tatsukura, Harada Shūsaku, and Shimizu Tsutomu, "Ichi gyū eki sei no hatarisu kansen ni shū te," *Hūten jūeki kenkyūjo kenkyū hōkoku* 1, no. 1 (1930): 123–40 and 221–22.

9. Kasai, "Cattle Epidemics in Manchuria and Mongolia and How to Prevent Them: Mission of the New Cattle Epidemiological Institute of the SMR Co., Mukden," *MDN*, 1 January 1926.

10. Bovine pleuropneumonia, or "lung plague," is caused by the bacterium *Mycoplasma mycoides mycoides*.

11. Both sheep pox and goat pox are caused by a virus from the family Poxiviridae and the genus *Capripoxvirus*.

12. *MDN*, 27 January 1914.

13. On the issue of defining the concepts of "modernity" and "colonial modernity" in East Asia and specifically on how such terms can be applied to an analysis of Meiji-Showa Japan and its empire (including Manchuria), see Sharon A. Minichiello, ed., *Japan's Competing Modernities: Issues in Culture and Democracy, 1900–1930* (Honolulu: University of Hawai'i Press, 1998); Gi-Wook Shin and Michael Robinson, eds., *Colonial Modernity in Colonial Korea* (Cambridge, MA: Harvard University Asia Centre, 1999); and Louise Young, *Japan's Total Empire: Manchuria and the Culture of Wartime Imperialism* (Berkeley: University of California Press, 1998), 241–43.

14. For a review of the history of Japanese imperialism during the Meiji era, see Marius B. Jansen, "Japanese Imperialism: Late Meiji Perspectives," in *The Japanese Colonial Empire*, ed. Ramon H. Myers and Mark R. Peattie (Princeton, NJ: Princeton University Press, 1984), 61–79.

15. On the history of Japan's creation and development of the puppet state of Manzhouguo, see Mark R. Peattie, *Ishiwara Kanji and Japan's Confrontation with the West* (Princeton, NJ: Princeton University Press, 1975); and Young, *Japan's Total Empire*.

16. On the issue of Manchuria as a contested space, see Owen Lattimore, *Manchuria: Cradle of Conflict* (New York: Macmillan, 1932); and Prasenjit Duara, *Sovereignty and Authenticity: Manchukuo and the East Asia Modern* (Lanham, MD: Rowman and Littlefield, 2003), 41–86.

17. Gavan McCormack, *Chang Tso-lin in Northeast China, 1911–1928: China, Japan, and the Manchurian Idea* (Stanford, CA: Stanford University Press, 1977), 4–5.

18. On the geographic or cartographic creations of Manchuria during the early twentieth century, see Sun Kungtu, *The Economic Development of Manchuria in the First Half of the Twentieth Century* (Cambridge, MA: East Asian Research Center, Harvard University, 1973), 41.

19. E. B. Schumpeter, "Japan, Korea and Manchukuo, 1936–1940," in *The Industrialization of Japan and Manchukuo, 1930–1940: Population, Raw Materials*

and Industry, ed. E. B. Schumpeter (New York: Macmillan, 1940), 299. The total area of Manzhouguo after the reorganization of the mid-1930s was 1,303,143 km², compared with 382,545 km² for Japan, and 675,377 km² for the rest of the empire.

20. Young, *Japan's Total Empire*, 40–46 and 55–114; and Rana Mitter, "Manchuria in Mind: Press, Propaganda, and Northeast China in the Age of Empire, 1930–1937," in *Crossed Histories: Manchuria in the Age of Empire*, ed. Mariko Asana Tamanoi (Honolulu: University of Hawai'i Press, 2005), 25–52.

21. On the career and thoughts of Gotō Shinpei, see Tsurumi Tasukuho, *Gotō Shinpei* (Tokyo: Kubisō shoten, 1985); Edward I. Chen, "Gotō Shinpei: Japan's Colonial Administrator in Taiwan; A Critical Re-examination," *American Asian Review* 13, no. 1 (1995): 29–59; and Liu Shiyong, "Medical Reform in Colonial Taiwan" (PhD diss., University of Pittsburgh, 2000), 65–79.

22. Liu, "Medical Reform in Colonial Taiwan," 74–75; Kitaoka Shin'ichi, *Gotō Shinpei: Gaikō to bijon* (Tokyo: Chūō koronsha, 1988), 60–65; Ruth Rogaski, *Hygienic Modernity: Meanings of Health and Disease in Treaty-Port China* (Berkeley: University of California Press, 2004), 258–59.

23. Yoshihisa Tak Matsusaka, *The Making of Japanese Manchuria, 1904–1932* (Cambridge, MA: Harvard University Asia Center, 2001), 90–92.

24. Kenichiro Hirano, "The Japanese in Manchuria, 1906–1931: A Study of the Historical Background of Manchukuo" (PhD diss., Harvard University, 1982), 241–45; and Matsusaka, *Making of Japanese Manchuria, 1904–1932*, 126–39.

25. John Young, *The Research Activities of the South Manchuria Railway Company, 1907–1945: A History and Bibliography* (New York: East Asian Institute, Columbia University, 1966), 4–6, 129–84, and 192–204.

26. Hirano, "Japanese in Manchuria," 244.

27. The plague that erupted in Manchuria in the autumn of 1910 was the pneumonic form of plague caused by the bacterium *Yersinia pestis.*

28. The term "plague-land" is a phrase borrowed from Ruth Rogaski, "Writing Western Hygiene in Nineteenth-Century East Asia," (paper presented at the Association for Asian Studies Conference, Washington, DC, 4–7 April 2003), 3–4.

29. On the history of the 1910–1911 Manchuria epidemic, see Kantō totokugu rinji bōrekibu, *Meiji yonjū-sen yonen minanmi Manshū, pesuto ryukō shifuroku* (Dairen: Manshū hibi shinbunsha, 1912); Iijima Wataru, *Pesuto to kindai Ch koku* (Tokyo: Kenbun shuppan, 2000), 137–74; Wu Lien-teh, *Plague Fighter: The Autobiography of a Modern Chinese Physician* (Cambridge: W. Heffer and Sons, 1959), 1–38; Carl F. Nathan, *Plague Prevention and Politics in Manchuria, 1910–1931* (Cambridge, MA: East Asian Research Center, Harvard University, 1967), 1–41.

30. Wu, *Plague Fighter,* 1–38; Wu Lien-teh, ed., *North Manchurian Plague Prevention Service: Reports (1911–1913)* (Cambridge: Cambridge University Press, 1914), 32–44; K. Kurauchi, "The Wild Rodents of Inner Mongolia: Plague Studies," in *Results of Work by the Hygienic Institute, SMR Company,* ed. Shōji Kanai (Dairen: South Manchuria Railway, 1931), 46–47.

31. Glanders is a disease caused by the bacterium *Burkholderia mallei.*

32. Anthrax is a disease caused by the bacterium *Bacillus anthracis.*

33. Matsuo, *Development of Science and Culture in Manchuria,* 6–9; Shōji Kanai, *Results of Work by the Hygienic Institute, SMR Company,* vol. 1 (Dairen: South Manchuria Railway, 1926), 98; and Manshikai, ed., *Manshū kaihatsu yonjūnen shi,* vol. 2 (Tokyo: Manshikai, 1964), 817–19.

34. See Robert John Perrins, "Doctors, Disease, and Development: Engineering Colonial Public Health in Southern Manchuria, 1905–1926," in *Building a Modern Japan: Science, Technology, and Medicine in the Meiji Era and Beyond,* ed. Morris Low (London: Palgrave Macmillan, 2005), 106–7 and 120–22; *KGFW* (1934), 93–96 and 121–22; Adachi Kinnosuke, *Manchuria: A Survey* (New York: Robert M. McBride, 1925), 134–39; and Abe Isamu, *The Economic Development of Manchuria: Japan's Contributions* (Dairen: Research Committee of the Pacific Relations in the South Manchuria Railway Company, 1931), 35–37.

35. Itō Takeo, *Life along the South Manchuria Railway: The Memoirs of Itō Takeo,* trans. Joshua A. Fogel (Armonk, NY: M. E. Sharpe, 1988), 24–28; and *Kantō-kyoku shisei sanjūnen shi,* 370–73.

36. *ROP* (1929), 120.

37. Itō, *Life along the South Manchuria Railway,* 27; *ROP* (1929), 120.

38. *KGFW* (1934), 121.

39. *ROP* (1931), 160–61.

40. *ROP* (1932), 146.

41. *KGFW* (1934), 121; and Adachi, *Manchuria: A Survey,* 139.

42. *ROP* (1931), 160–62.

43. Abe, *Economic Development of Manchuria,* 36–37; and Matsuo, *Development of Science and Culture in Manchuria,* 3–4.

44. *Economic History of Manchuria* (Seoul: Bank of Chosen, 1921), 159–60; Henry W. Kinney, *Manchuria Today* (Dairen: Hamada, 1930), 62–64.

45. For example, see Henry W. Kinney, *Manchuria: Land of Opportunities* (New York: South Manchuria Railway, 1922), 13; and Henry W. Kinney, *Modern Manchuria and the South Manchuria Railway Company* (Dairen: South Manchuria Railway Company, 1928), 58.

46. Figures compiled from *ROP* (1931), 102–30.

47. See Hirano, "Japanese in Manchuria," 244.

48. *KGFW* (1934), 106–7.

49. *Manchuria: Land of Opportunities,* 15; Adachi, *Manchuria: A Survey,* 172–78.

50. *ROP* (1929), 121; also cited in Matsuo, *Development of Science and Culture in Manchuria,* 3.

51. L. S. Palen, "The Romance of the Soya Bean," *Asia* 19, no. 1 (1919): 68–74; *ROP* (1931), 161.

52. *Light of Manchuria* (July 1922), 1–12.

53. Nakano Seiichi, *Manshū no chikusan* (Tokyo: Meibundō, 1942), 5–8; *Manchuria: Land of Opportunities,* 25–26.

54. Y. Sakatani, *Manchuria: A Survey of Its Economic Development,* rev. Grover Clark (1932; rev. New York: Garland, 1980), 260.

55. Kawamura Suga, ed., *Manshū chikusan gaiyō* (Shinkyō: Manshū jijō annaijo, 1940), 1–2; and Abe, *Economic Development of Manchuria,* 37.

56. Sakatani, *Manchuria: A Survey of Its Economic Development,* 260–62; *KGFW* (1934), 122; and Nakano, *Manshū no chikusan,* 21–27, 40–49, and 53–61.

57. *ROP* (1936), 87–89; and Manshikai, *Manshū kaihatsu yonjūnen shi,* vol. 2, 818–22.

58. Kinney, *Modern Manchuria and the South Manchuria Railway Company,* 61.

59. *Manchuria: Land of Opportunities,* 26; Sakatani, *Manchuria: A Survey of Its Economic Development,* 261–62.

60. Abe, *Economic Development of Manchuria,* 37.

61. Sanga Isao, *Manshū no nō chikusangyō no gensei* (Dairen: Manshūbunka kyōkai, 1932), 6; and *ROP* (1936), 87–89.

62. Sun, *Economic Development of Manchuria in the First Half of the Twentieth Century,* 49; Kang Chao, *Economic Development of Manchuria: The Rise of a Frontier Economy* (Ann Arbor: Center for Chinese Studies, University of Michigan, 1982), 50; and *Manshū nenkan* (Dairen: Manshū nichinichi shinbun, 1942).

63. See Sekiya Osamu, *Manshū bokujō ki* (Hōten: Hōten daihanyugō shoten, 1944), 250–68; Matsumoto Toshio, *Tetsudō sōkyoku chikusan jigyō hōsaku uchiwasekaigi giji hōkoku* (Dairen: Mantetsu chōsabu, 1939), 1–2; and *Manshūkoku zenshō shokusan kachō narabini chikusan shuninkan kaigi hōkoku* (Dairen: Mantetsu chōsabu, 1939), 1–3 and 33.

64. See Ian Clunies Ross, *A Survey of the Sheep and Wool Industry in North-Eastern Asia: With Special Reference to Manchukuo, Korea and Japan* (Melbourne: Australian Council for Scientific and Industrial Research, 1936), 10–25.

65. Kinney, *Manchuria Today,* 64.

66. For example, see *ROP* (1929), 139–40; and Adachi, *Manchuria: A Survey,* 179–80.

67. *ROP* (1936), 129.

68. On the history of Japan's settlement programs in Manchuria after 1931, see Young, *Japan's Total Empire*, 307–98.

69. *ROP* (1939), 117.

70. Young, *Japan's Total Empire*, 321–22.

71. Takumushō takumukyoku, "Hokuman ni okeru shūdan nōgyō imin no keiei hyōjun'an," *Nagai chōsa shiryō* 10, no. 11 (1938): 11–13.

72. For an overview of the history of Manzhouguo's Five-Year Plans, see Ramon H. Myers, "Creating a Modern Enclave Economy: The Economic Integration of Japan, Manchuria, and North China, 1932–1945," in *The Japanese Wartime Empire, 1931–1945*, ed. Peter Duus, Ramon H. Myers, and Mark R. Peattie (Princeton, NJ: Princeton University Press, 1996), 151–59.

73. *ROP* (1939), 58–81.

74. *ROP* (1939), 62; Schumpeter, "Japan, Korea and Manchukuo, 1936–1940," 307–8.

75. Kihara Rinji and Uchigasaki Kenjirō, *Nōji shisetsu oyobi nōji gyōseki* (Dairen: Mantetsu chōsabu, 1938), 94–95.

76. *Manshūkoku shi*, vol. 1 (Tokyo: Manshūkoku shi hensan kankō kaisan, 1972), 719–20 and 726–27.

77. For a list of these rural reports, see Young, *Research Activities of the South Manchurian Railway Company*, 129–84 and 192–204.

78. Nakano, *Manshū no chikusan*, 1–8 and 35–39.

79. Chao, *Economic Development of Manchuria*, 51.

80. For information on anthrax outbreaks in Manchuria, see Sekiya, *Manshū bokuj ki*, 85–89. For accounts of epidemics of rinderpest, sheep pox, and glanders during the late 1930s, see MDN, 15–29 June 1938, 19 November 1938, and 19 March 1939.

81. *ROP* (1929), 119–22; Kanai, *Results of Work by the Hygienic Institute*, 95–98.

82. See Director Kasai Katsuhiro's preface to *Hōten jūeki kenkyūjo kenkyū hōkoku*, vol. 1 (1930), 1–3; *ROP* (1931), 228–29; *MDN*, 29 December 1925.

83. Foot-and-mouth disease (FMD) is caused by an apthovirus from the family Picornaviridae.

84. Hog cholera, or swine fever, is caused by a virus from the family Flaviviridae and the genus *Pestivirus*.

85. Rabies is caused by a *Lyssavirus* in the family Rhabdovirus.

86. *ROP* (1934), 241–42.

87. Kasai, "Cattle Epidemics in Manchuria and Mongolia and How to Prevent Them," *MDN*, 1 January 1926.

88. *ROP* (1931), 228–29.

Chapter 12

Sheep Breeding in Colonial Canterbury (New Zealand)

A Practical Response to the Challenges of Disease and Economic Change, 1850–1914

Robert Peden

SHEEP FARMING was the most important agricultural industry in New Zealand from the 1850s to late in the twentieth century. The industry was founded on fine-wooled Merinos imported from the Australian colonies, and wool was the most valuable single agricultural export until the late 1960s.[1] However, the year 1882, when the first shipment of frozen mutton was made from New Zealand to Great Britain, marked a watershed. From that time, meat became increasingly important to New Zealand's economy. The crossbreeding of sheep, and in particular the development of the Corriedale breed in colonial Canterbury, is often seen in the context of the establishment and expansion of the frozen-meat industry.[2] The argument, put simply, is that, before the advent of refrigeration, farmers raised Merino sheep for their wool; after 1882, they raised crossbred sheep for their wool *and* their meat. However, experiments crossing British rams over the base Merino flock began almost from the outset of organized settlement and were a response initially to the disease of footrot.

There were also economic conditions in the prerefrigeration era that encouraged sheep breeders to experiment with crossbreeding—in particular, the increasing demand for combing wool by English processors and the requirement for a larger-framed, meatier, and faster-maturing type of sheep than the Merino for the local butchers' market and for boiling down. In fact, it was the very success of these experiments in crossbreeding that enabled Canterbury farmers to take advantage of the new market made available by the opening of the frozen-meat trade with Great Britain after 1882. In this chapter, I examine these issues and explore the dilemma that breeders faced in trying to develop a sheep breed that was suited to the local environment, one that would resist the challenges of footrot and at the same time satisfy the changed demands in the marketplace for wool and surplus sheep.

I contend that footrot was a primary reason for the early experiments in crossbreeding. Footrot was the most serious sheep disease in the Canterbury region from the late 1860s, and it continues to be a major animal-health concern for present-day farmers who raise fine-wooled sheep. A 2001 survey of New Zealand Merino farmers cited footrot as the second most significant disease after gastrointestinal parasitism.[3] Footrot has significant economic costs for farmers. Sheep become lame and are less inclined to graze, with the result that they lose weight, grow less wool, have a lower lambing performance, and are more prone to fly strike, which, if not treated, will lead to their slow and painful deaths. Over and above the costs from the loss of production, the management of the disease is expensive in terms of the cost of treatment and the cost of labor.

If the disease is such a problem for Merino farmers in the twenty-first century—when we know its etiology and have a sound understanding of its management and when farmers have access to vaccines and antibiotics—then pity the farmer in the nineteenth century who had none of this knowledge and only limited scientific and veterinary support.

Etiology of Virulent Footrot

To set a context for the problem of footrot in the colonial setting, we should first look at the etiology of the disease. Footrot is caused by the combined effect of two gram-negative anaerobic bacteria: *Fusobacterium necrophorum* and *Dichelobacter nodosus*.[4] In warm, moist conditions, the skin between the claws of sheep can become softened and raw, allowing the hoof to be invaded by these bacteria. Epidermal penetration by *F. necrophorum* creates the condition of ovine interdigital dermatitis. The presence of *D. nodosus* at this stage results in virulent footrot. As the disease spreads in the hoof, it destroys infected tissue so that the horny part can become almost completely detached. Virulent footrot is highly contagious and can result in 100 percent morbidity.[5]

D. nodosus is able to survive for long periods within the hoof with no external signs that the hoof is infected, but it does not survive outside the host for more than two weeks. The transmission of infection is determined by environmental conditions, and moist conditions above ten degrees centigrade are a precondition for the development and spread of the disease. Increased stocking rates will increase the rate of disease spread. A long spell of dry conditions will result in a spontaneous cure, but some sheep will become hosts and carry the bacteria for years and on into the next challenge season.[6]

The treatment of the disease is labor-intensive and slow. Each hoof has to be inspected. Where there is any external sign of infection, the hoof needs to be pared away carefully to expose the diseased tissue. This exposes the anaerobic *D. nodosus* to the air, which helps in the control process. It also exposes it to any topical bactericidal agents that might be applied, but their application may have to be repeated several times to achieve a cure.[7] There are also vaccines and antibiotics that will cure the disease, but these are expensive.

All sheep breeds are susceptible to footrot, but Merinos are more prone than other breeds. Virulent strains of the disease are more pathogenic in Merinos than in British breeds and, after field trials, J. R. Egerton and others have reported that "the incidence, severity, duration, and extent of infection was higher among Merinos than in Border Leicester-Merino crossbreds."[8] Merino sheep originated in the semi-arid regions of the Mediterranean and were not exposed to the challenge of footrot, so did not develop resistance to the disease.

Footrot in Colonial Canterbury

Organized settlement began in Canterbury in 1850, and within a short time, large numbers of sheep were being imported from the Australian colonies of New South Wales and the Port Phillip District (now Victoria). There is no doubt that footrot arrived with those sheep. Despite Australia's reputation for aridity, footrot was a serious disease in its colonial era and, in fact, remains so today. The Australian sheep flock was predominantly Merino and, therefore, highly susceptible to the disease. William Youatt wrote in 1837 that "footrot seems to assume a character of its own in New South Wales . . . [and] if neglected, it speedily becomes inveterate, and preys upon and destroys the animal. The losses occasioned by it in the early existence of the colony were frightful."[9] Alfred Joyce, a squatter in the Port Phillip district, complained that footrot and scab cost him three thousand pounds in 1853 alone—this being made up of stock losses and the cost of treatment.[10]

In New Zealand, footrot was a problem from the outset in wetter regions in the North Island, although in Canterbury the disease does not appear to have been a problem on the large sheep stations, perhaps because of the low sheep numbers in the early years and the management system where the sheep were not confined but allowed to run on open blocks, which meant that the disease would not have been readily transmitted.

However, settlers on their small farms around the towns and villages found Merinos poorly adapted to their heavier land. The Deans brothers, who ran sheep on the Canterbury Plains before organized settlement began, noted how quickly the hooves of their Merinos grew on the heavy country at their Riccarton farm. They decided that sheep farming might be more successful on the hill country to the west, where the sheep's hooves would be kept shorter on the stonier ground.[11] By 1860, farmers were experimenting with crossbreeding, using sires of English breeds over Merino ewes. Cotswold, Hampshire, and Southdown rams were noted in newspaper advertisements and articles.[12] In 1866, there were more Leicesters, Cheviots, and Romney Marsh sheep shown at the Canterbury Agricultural and Pastoral Association Exhibition than there were Merinos.[13]

Pastoralists on the stations of the open plains and mountain lands continued to raise Merino sheep, but changing conditions that encouraged the spread of footrot began to have an impact on their farming operations. Sheep numbers built up remarkably quickly, so that by the end of the 1860s, the country in its native state was said to be fully stocked. From about the same time, the use of wire fencing and large-scale cultivation intensified on the Canterbury Plains, which led to sheep being run on increasingly confined blocks.

The use of turnips and replacing the native vegetation with introduced grasses and legumes combined to increase the stocking rates from perhaps one sheep to three acres to one sheep to the acre or better. This provided ideal conditions for the spread of footrot. A correspondent to the *New Zealand Country Journal* in 1879 emphasized this when he wrote: "It is impossible to keep the Merino on the bulk of our cultivated lands, for the simple and best of reasons—their feet are not adapted to moist lands, and that footrot is the result."[14] William Soltau Davidson, manager of The Levels station and one of the breeders instrumental in developing the Corriedale, wrote in his memoir: "The introduction of English grass pasturage necessitated a change in the stocking of the properties, because merino sheep when grazed on cultivated land soon became afflicted with foot-rot. It was therefore necessary to adopt a long-wooled breed and their crosses, which throve well on the English grasses."[15]

Table 12.1. The expansion in sheep numbers, acres fenced,
and acres under crop in Canterbury between 1855 and 1881

Year	Sheep	Acres fenced	Under crop
1855	220,000	12,200	6,460
1861	877,369	72,900	32,800
1864	1,500,000	217,000	50,000
1867	2,500,000	1,000,000	N.A.
1874	3,325,000	N.A.	470,300
1881	3,520,000	4,150,000	1,000,000

Source: Compiled from Census 1864–65, published in the Lyttelton Times, January 18, 1865, 2; Statistics of New Zealand, 1869, 1882 (Wellington: Government Printer); B. L. Evans, Agricultural and Pastoral Statistics of New Zealand 1861–1954 (Wellington: Government Printer, 1965).

Even on the hill and high-country runs, footrot became a problem as some of the more progressive pastoralists engaged in large-scale improvement programs.[16] For example, by 1871, Orari Gorge station had put up thirty miles of fencing; three years later, this had been extended to sixty miles. In combination with a cultivation program on the flats and terraces, it enabled the station to increase its sheep numbers to forty thousand by 1879.[17]

Te Waimate was another property that embarked on a large development scheme of drainage, cultivation, and fencing that led to increased sheep numbers, but also to devastating outbreaks of footrot. Owner E. C. Studholme described the tedious job of treating sixteen thousand infected sheep in one winter.[18] The station staff spent most of the winter months trying to cure the sheep by paring their feet and then running them through a trough containing a mixture of arsenic, bluestone, soda, and water. Studholme gives us some insight into the labor-intensive nature of footrotting sheep, writing "footrot was a horribly monotonous job after one had been at it for weeks, and it was a great relief when the trouble disappeared for the time being."

Mount Peel station was the first high-country station taken up in Canterbury. The runholder John Barton Acland was a progressive farmer and instigated a cultivation and fencing program that began almost as soon as the station was established in 1856. The flats and terraces in the vicinity of the homestead were plowed and sown in turnips for wintering hoggets and later for fattening sheep for market, or sown into pasture for hay or grazing. Inevitably, this intensification resulted in severe outbreaks of footrot in the Merino flock in wet seasons.

The farm diaries from Mount Peel are not detailed enough to provide any idea of actual man-hours involved in treating footrot, but they do offer

Table 12.2. Days recorded treating footrot at Mount Peel, 1880–94

Year	Days
1880	16
1881	2
1882[1]	17
1883	45
1884	31
1885	34
1886	11
1887	15
1888	10
1889[2]	Nil
1890[2]	Nil
1891[2]	1
1892	30
1893	46
1894	12

1. There are no diary entries from March 21 to May 2. This was normally a busy time for footrotting: checking sheep after weaning and preparing the rams and ewes for mating. In 1883, footrotting took place on thirteen days over the same period.

2. These were dry years: 1889, Jan.–Aug.: 19 inches, and Sept.–Dec.: 12 rain days; 1890, 38.71 inches; 1891, 41.32 inches. Average rainfall at Mount Peel: 45 inches.

Source: MB44, Acland Family Papers, boxes 4, 51, 52, Macmillan Brown Library Archives, University of Canterbury.

some idea of the extent of the problem. In 1873, the diary noted that the rams were being treated for footrot, and from that time, the disease was ever present in the ram flock.[19] It soon spread to other classes of sheep on the station. Table 12.2 shows that footrot was an ongoing problem through the 1880s, although the disease went into remission with the onset of a series of dry years in 1889, 1890, and 1891. However, a wet year in 1892 launched another outbreak.

The treatment of footrot on Mount Peel seems consistent with other contemporary accounts dealing with the problem. The sheep were inspected and lame animals isolated. Their feet were pared to remove the infected tissue, and then the animals were run through a trough containing a solution of arsenic and bluestone.[20] The infected sheep were brought in regularly to have their feet "dressed"; but despite these measures, the cure rate for footrot was low, and the problem was only overcome temporarily in dry seasons when the disease went into remission naturally.

So, the problem of footrot was a very real issue for pastoralists and farmers in colonial Canterbury. The national flock was based on footrot-prone Merino sheep. The intensification of farming methods and increasing stock numbers forced sheep into ever-closer contact and spread the infection. Despite their best efforts to cure their infected animals, sheep farmers found the problem insurmountable in all but the driest years. Naturally, they looked for a long-term solution, and that was to change their sheep breed. Yet, this was a highly contentious issue in the region from at least the early 1860s. However, before going on to scrutinize that debate, we need to examine two other factors that encouraged sheep breeders in Canterbury to experiment in crossbreeding.

Changing Demand for Wool

High prices for fine Merino wool, which was sought after by English, European, and American woolen mills, drove the expansion of the pastoral industry in New Zealand. However, from the late 1860s, wool prices began a decline that continued for the remainder of the century, and very fine Merino wool, which had been in high demand, for a time fell out of favor. Structural changes in the wool-processing industry in England and changes in consumer fashion in Great Britain and Europe led to the expansion of the worsted industry centered in the West Riding of Yorkshire. Worsted processors wanted "combing" wool. This has a long staple and good tensile strength to withstand mechanical combing without breaking. English wool growers were unable to meet the increasing demand for this type of wool, so the processors turned to the colonies for their fiber. Reports from the London wool sales in the *Lyttelton Times* from early 1860 began to stress that the worsted districts demanded "sound, shafty" combing wool.[21] An article in the *Mark Lane Express*, reprinted in Christchurch in 1864, argued that

> all the new sources of supply—Australia, Tasmania, South Africa, New Zealand—furnish fine, soft, useful, short-stapled wool [while] the demand for long-grown wool increases year by year, and any country which possesses facilities for the production of a wool endowed with qualities which are peculiar to wool of English growth seems far more likely to ensure a profitable market for its commodity than to adhering to wool of a shorter and finer type.[22]

Businessmen from Bradford and Halifax formed the Wool Supply Association to encourage wool growers to change to long-wooled breeds. The association made submissions to governments and sent letters and pamphlets to newspapers and periodicals as part of its promotional campaign. In December 1863, the *Lyttelton Times* published a letter from the secretary of the Bradford Chamber of Commerce, who saw New Zealand's future as a supplier of "long-stapled fleeces, of a medium quality and length, between the fine Merinos of Australia and the long-grown Leicester of [England]."[23] The letter continued that if New Zealand farmers used English breeds instead of pure Merinos, they would "produce a fleece better adapted to meet the growing wants of the manufactures of this country."

Naturally, prices on the auction floors in London backed up this pressure on growers to change the wool type that they produced. A compelling example of the price advantage of combing wool over fine Merino can be found in prices received by The Levels station. In 1868, Leicester rams were used over a line of Merino ewes; hogget wool from the resulting cross sold for sixteen pence per pound, whereas Merino wool from the station fetched nine pence.[24] In addition to the price advantage of the halfbred wool was the weight advantage that it gave over wool cut from pure Merinos. Holme Station in South Canterbury had a highly regarded Merino flock, yet in 1872, hoggets bred from the cross of a Leicester ram over Merino ewes cut nearly seven pounds of wool per head, whereas the Merino ewes clipped five and a half pounds.[25]

In 1871, the *Timaru Herald* published a letter from an unnamed London wool broker that clearly laid out the advantages of halfbred sheep over Merinos: "The demand for half-breed wools at enhanced prices has attracted considerable notice, and is almost certain to continue . . . and when the weight of fleece and carcase is considered it does not require much more to prove that this description of sheep will be much more paying than the Merino."[26] Clearly, from the 1860s, there was an economic advantage in producing wool from halfbred sheep. However, there was a spirited debate in the region over the practicalities of this. But before exploring that debate, I want to look at the third advantage to be gained in crossbreeding.

Sheep Fattening

Merino sheep were unsuited for meat production. For centuries they had been bred only for the quality of their wool. They were small framed, lean, and slow to mature compared to British breeds like the improved Leicester and Lincoln. In Great Britain, Merinos had been fashionable around the turn of the nineteenth century, but because of these very problems and their proneness to footrot, they quickly fell out of favor with English farmers.

As noted earlier, farmers on the heavier country near Canterbury's settlements had begun experimenting with crossbreeding by at least the beginning of the 1860s and this was, in part, to overcome the problem of footrot. Another reason was that they relied not only on wool for their income but also on producing food for the local townships, and Merino sheep were simply not productive enough. In 1861, Cotswold-Merino ram lambs were advertised with the claim that in Australia this cross had been found to "improve the constitution, to increase and ripen the carcase, add greatly to the weight of wool and length and strength of staple."[27] At the Pastoral and Agricultural Show held in 1862, there were two classes for fat sheep: Robert Chapman won the class where sheep were required to be fattened on native pasture with a line of four-year-old Merino wethers, with the heaviest carcass weighing eighty pounds; Mrs. Deans won the section where sheep were finished on enclosed ground with eighty pound Merino-Southdown halfbred wethers that were only ten months old.[28] This proved to local farmers that the advantage of crossbreeding for meat production was beyond dispute.

By the end of the 1860s, the local demand for surplus sheep off the runs disappeared. Up to this time, there had been a ready market for sheep as pastoralism expanded, but by 1865 all the country suited to extensive sheep farming had been taken up; by the end of the decade, the runs, in their native state, were fully stocked. As gold petered out, the miners drifted away from the goldfields of Otago and Westland and with them went a large market for meat. Apart from the butchers' market in the region's towns, the only outlets for surplus sheep were the boiling-down plants where sheep were rendered into tallow for export. As with meat production, the lean and slow-maturing Merinos did not have the size or condition to make rendering down profitable. As early as 1870, the *Timaru Herald* reported that halfbred sheep were being expressly bred for boiling down and meat preservation.[29] In 1872, a commission agent, who bought and sold stock on behalf of farmers, reported that it was evident from the number of long-wooled rams being sold that stockowners were going in largely for crossbreds.[30] Later in the same year a newspaper report noted that the "rapidly increasing area of land under English grass has convinced our farmers of the necessity of turning their attention to breeds other than the merino."[31]

The Great Merino Debate

It would seem that the advantages in crossbreeding over persevering with the pure Merino were compelling. Merinos were prone to footrot, their wool was becoming less competitive in the marketplace, and they were not

suited for meat production or for boiling down. Yet the idea of crossbreeding produced intense debate in the region, with enthusiasts for the Merino
breed predicting disaster for the future of the sheep industry if the practice
became widespread. After Mrs. Deans's success at the Pastoral and Agricultural Show with her Merino-Southdown halfbreds, a writer in the *Lyttelton
Times* warned:

> The high price of butcher's meat at present is an inducement
> to speculate in [crossing], and the home demand for a longer
> stapled wool is also a stimulant to experiments. The sheep
> farmer should pause before embarking in such an experiment,
> as the practice is not sustained in theory. The first cross of the
> Merino with the South Down may yield a sheep well adapted for
> the purpose, but breeding from such would entail degeneracy
> and a mongrel race.[32]

A year earlier the *Lyttelton Times* had reprinted an article from *The Spec-
tator* that questioned the wisdom of New Zealand sheep breeders' shifting
the focus of their production from short, fine wool to long, stronger wool.
It asserted that it was impossible to establish a breed of sheep by crossing
two breeds and obtaining the merits of both. The article quoted a paper by
William Charles Spooner, M.R.V.C., that had been published in the *Econo-
mist* where he emphasized: "Cross-breeding is merely a plan of producing
meat, for cross bred animals are only profitable when bred for the butcher.
They cannot be perpetuated."[33] Spooner went on to say that no one should
cross to establish a new breed "unless he has clear and well defined views
of the object he seeks to accomplish, and has duly studied the principles
on which it can be carried out, and is determined to bestow for the space
of half a lifetime his constant and unremitting attention to the discovery
and removal of defects." The article outlined another objection to crossbreeding: the resulting lack of uniformity in the character of the wool. It
stressed the maxim "like produces like" and claimed that crossing produces
"innumerable varieties, and not infrequently on the same sheep."

Another article with a similar message was reprinted in the *Lyttelton
Times* in 1863. It was written by a Professor Ran and had been published in
the Hohenheim weekly paper. Professor Ran was referred to as "one of the
first authorities in Europe on the subject of the weight of fleece and carcase
of sheep," and his objection to crossbreeding was that it was difficult to
obtain a large carcass and an abundance of wool together.[34] Wool growth,
he claimed, occurred at the expense of carcass growth, and farmers who

were concerned with wool production should breed Merino sheep with a small-to-average frame, while those who wished to grow sheep with a large carcass must expect to grow less wool.

Thus, we have the dilemma of sheep owners in colonial Canterbury: the market and the problem of footrot encouraged them to change their type of sheep, while the theorists of the time told them it could not be done without risking the quality of their flocks. Many tried to establish a two-flock system where Merino rams were used over part of the Merino ewe flock to maintain the base breed and rams of British breeds were crossed over the rest of the ewes. The progeny of this cross was known as a halfbred, and it produced ideal wool for the worsted trade; the wethers fattened quickly and grew to heavy weights, but what to do with the ewes from the cross? Nor did this really deal with the footrot problem as the base breeding flock remained Merino.

Experiments in Crossbreeding

At Mount Peel station, John Barton Acland remained a committed Merino man despite the problems of footrot. Yet by 1884 even he relented and began using Leicester rams over some of his Merino ewes. By 1896, the hill flock remained pure Merino, but nearly half was mated to English Leicester rams. Acland died in 1904, and the new manager moved quickly to change the breed of the base flock. In 1911, only 476 of the 13,057 lambs marked were Merino.[35] While Mount Peel was slow to change, others reacted much more quickly. In 1875, Clent Hills station, a run in the Ashburton Gorge, was advertised for sale with 19,500 sheep of which 15,628 were crossbred.[36] The Levels station mated a cut of their Merino ewes to Leicester rams in 1868 as an experiment. The results were so successful that by 1879, of the 79,497 sheep on the run, only 6,300, less than 8 percent, were pure Merinos.[37]

These examples show that sheep breeders were making the change away from the early reliance on the pure Merino, but the problem of how far to go with the crossing remained unresolved. There was, however, a clear awareness that different breeds of sheep suited different environments. An article on the crossbreeding of sheep in the *New Zealand Country Journal* in 1877 expressed this very point, noting the "need to breed sheep to suit the country."[38] The writer went on to say that the Merino was best suited to hilly country; the first cross Leicester-Merino halfbred was best for English grass pasturage; and for heavy, low-lying ground, a three-quarter-bred sheep, the product of using a Lincoln ram over the Leicester-Merino half-bred, was the ideal type. In the same issue of the journal, another writer emphasized the importance of matching the sheep to the environment and argued that the Leicester-Merino halfbred was the best type for the drier portions of the Canterbury plains.[39]

Acland tried to keep the Merino as the base breed on his mountain lands but ran first-cross halfbreds on his improved easier country. At Clent Hills, the Merino-Leicester cross-ewes were mated to Lincoln rams, thereby moving them further away from the Merino. At The Levels, they tried a different approach. William Soltau Davidson, who managed the station, described the crossbreeding problem:

> While a supply of Merino ewes was available, excellent half-bred sheep were easily enough bred by crossing them with Lincoln or Leicester rams, but it was the after-breeding that was the difficulty. If the half-bred ewes were mated with Longwool rams the progeny—three-quarter-breds as we called them—were heavier sheep than we desired; while if merino rams were used the progeny were too small and were uneven in the wool. It was the half-bred sheep we wanted and nothing more or less.[40]

In an attempt to overcome these problems, in 1874 Davidson set out to breed an inbred halfbred that would breed true to type.[41] He joined selected stud Lincoln rams to one thousand ewes from The Levels Merino stud flock. The first mating produced about 450 ewe lambs, from which Davidson selected 150 for the breeding program. He went on to breed from the original parents until they became unproductive, while continuing the policy of heavily culling the progeny. In time, the young halfbred ewes were mated with rams chosen from their own lot; this strategy of inbreeding continued, so that forty-four years later Davidson was able to write that The Levels Corriedale stud flock contained "no other blood than that originally adopted to create the type, which is now absolutely fixed."[42]

The Corriedale

The person usually credited with establishing the Corriedale breed is James Little, a Scot who managed Corriedale station in Otago. He started cross-breeding experiments using Romney rams over six hundred Merino ewes in 1868.[43] The halfbred progeny were mated together, and this program was continued with success until Little moved to his own property, Allandale, in North Canterbury in 1878. There, he started again with two thousand selected Merino ewes, which he mated to Lincoln rams. From the progeny, he selected twenty ram lambs that met the type he desired. These were later mated with selected ewes from the same cross.[44]

As a result of the work of Little, Davidson, and others, the inbred half-bred became highly regarded throughout the low-rainfall districts of New Zealand. The success of these experiments was also noted elsewhere. An article from the British publication the *Live Stock Journal* was reprinted in the *New Zealand Country Journal* in 1883 and recognized both the intellectual and practical breakthrough that these breeders had made. It criticized sheep breeders in New South Wales because they had been so swayed by the belief that crossbreeding was doomed to fail that they had lacked "the courage to attempt any experiments outside . . . the merino branch of the sheep-breeding industry." The article went on to note:

> There are, however, a few who, not misled by mere theories, are persevering with sheep such as are now attracting much attention in New Zealand. The breeders of New South Wales have nought to offer English consumers but lean, small-carcased merinos, while New Zealanders are able to compete with even the Southdowns by sending to the London Market well-fattened cross-breds.[45]

The inbred halfbred was already widely known as the Corriedale in the 1890s, and this name was officially sanctioned in 1905 by the New Zealand Sheep Breeders' Association.[46] In 1910, Corriedale breeders in New Zealand formed their own association, and in 1916, twenty Corriedale flocks were admitted to the N.Z. Sheepbreeder's Association Flock Book.[47] From the outset, the Corriedale was bred to be both a wool- and meat-producing sheep. Depending on the objectives of the breeder, the Corriedale produces a medium-to-fine, long-stapled fleece with a well-defined crimp. It was a wool type that found a ready market in the worsted trade. The Corriedale is more fecund than the Merino, and the lambs mature early to produce a well-muscled carcass.

F. R. Marshall, from the United States Department of Agriculture, visited New Zealand in 1914 to assess whether the Corriedale should be imported into the United States. His description of the character of the country where the breed was run in New Zealand illustrates the range of environments to which the Corriedale is adapted. Marshall noted that it varied "from level and rich artificial grass pastures to rough hills with altitudes around 3,000 feet, on which snow sometimes lies for several months at a time."[48] On wetter, low-lying ground, the Corriedale remained susceptible to footrot, and the Romney Marsh breed became increasingly popular on this heavier country.

Marshall was impressed with both the wool and meat-producing quali-
ties of the Corriedale. He commented that while its breeders had given
"special emphasis" to wool, they had not ignored carcass development and
that Corriedale lambs were ready for the market at six months of age.[49] As
a result of his tour, fifty-three ewes and ten rams of the Corriedale breed
were shipped from New Zealand to San Francisco in December 1914.[50] The
Corriedale also became established in Australia, South Africa, and eastern
Europe; but it has been in Peru, Argentina, and Chile that it has become
easily the most popular breed.

THE FIRST shipment of frozen meat to Great Britain in 1882 has always
been viewed as pivotal in New Zealand's economic and agricultural history. The orthodox historical perspective of the new technology has tended
to focus on the influence of the market and to neglect the environment
and the sheep that enabled Canterbury farmers to take advantage of the
opportunity it provided. Farmers in New Zealand were innovative from
the outset of settlement, and both environmental and economic drivers
shaped their actions. In this chapter, I have set out to show that crossbreeding began very early in colonial Canterbury and was already an established
practice well before 1882.

It is my contention that the disease of footrot was a primary factor
behind the early experiments in crossbreeding. For small farmers on heavy
country, footrot was a problem from the beginning. On run country, it was
not an issue until the intensification of farming methods actually changed
the environment of the rangelands and the subsequent increase in stocking
rates led to footrot becoming endemic among Merino sheep on improved
pasturage. This problem, on its own, was enough to encourage sheep breeders to explore crossbreeding as a long-term solution to footrot. Yet, at the
same time, changing economic forces were also having an impact on the
profitability of Merino sheep. Changes in the type of wool required in the
English market place and the unsuitability of Merinos for the meat trade
were other factors that encouraged sheepmen in Canterbury to look to a
fixed inbred halfbred type to meet their requirements. Against the advice
of the contemporary experts, practical farmers set out to develop a new
breed of sheep.

The Corriedale was the result of their experiments. It proved to be
more productive than the Merino at higher stocking rates on improved
pastures and was suited to open plains and hill country. While it is far from
being footrot-resistant, the Corriedale is less susceptible to the disease than
are Merinos. The early breeders succeeded in developing a dual purpose

sheep with a heavy, fine-to-medium fleece and the ability to produce lambs that mature early and grow a meaty carcass. As a result of possessing these attributes, the Corriedale now rivals the Merino as the most numerous sheep breed in the world.

Notes

1. B. L. Evans, *A History of Agricultural Production and Marketing in New Zealand* (Palmerston North: Keeling and Mundy, 1969), 89.

2. Ibid., 5, 82; Corriedale Sheep Society, *The Corriedale, "New Zealand's Own Breed": History of Development* (Christchurch: Corriedale Sheep Society, 1939), 10; P. G. W. Stevens, *James Little and His Corriedale Sheep* (Christchurch: Canterbury Agricultural College, n.d.), 8; William Perry, *Sheep Farming in New Zealand* (Christchurch: Whitcombe and Tombs, 1923), 52.

3. Chris Mulvaney, *A Guide to the Management of Footrot in Sheep* (Wellington: Woolpro, 2002), 9.

4. *The Merck Veterinary Manual,* "Virulent Footrot," http://merckvetmanual.com/mvm/index.jsp?cfile=htm/bc/90906.htm&word=footrot (accessed 3 May 2007).

5. Ibid.

6. Mulvaney, *Guide to the Management of Footrot,* 16–19.

7. Ibid., 35, 37.

8. D. J. Stewart, "Footrot in Sheep," in *Footrot and Foot Abscess of Ruminants,* ed. J. R. Egerton, W. K. Yong, and G. G. Riffkin (Boca Raton, FL: CRC Press, 1989), 11.

9. William Youatt, *Sheep: Their Breeds, Management, and Diseases to Which Is Added the Mountain Shepherd's Manual* (London: Simplan, Marshall, 1866), 189.

10. G. F. James, ed., *A Homestead History, Being the Reminiscences and Letters of Alfred Joyce of Plaistow and Norwood, Port Phillip, 1843 to 1864* (Melbourne: Melbourne University Press, 1942), 139.

11. John Deans, ed., *Pioneers of Canterbury: Deans Family Letters, 1840–1854* (Dunedin: Reed, 1937), 135.

12. In the early years of crossbreeding, the terms *halfbred* and *crossbred* were used without distinction. However, as the practice became more widespread, the term *halfbred* was increasingly used only for the first cross between a Merino and a longwool, and the term *crossbred* was used for subsequent crosses, such as the three-quarter bred and beyond. A longwool described British breeds such as the Leicester, Lincoln, and Romney Marsh, in contrast to the short-wooled sheep like the Down breeds, such as the Southdown, which grew shorter and finer wool.

13. *Lyttelton Times,* 10 November 1866.

14. T. T., "To the Editor of the New Zealand Country Journal," *New Zealand Country Journal: A Record of Information Connected with Agricultural, Pastoral and Horticultural Pursuits and Rural Sports in New Zealand* 3, no. 6 (1879): 395.

15. W. S. Davidson, *William Soltau Davidson, 1846—1924: A Sketch of His Life Covering a Period of Fifty-two Years, 1846–1916, in the Employment of the New Zealand and Australian Land Company Limited* (Edinburgh: Oliver and Boyd, 1930), 23.

16. In the colonial period, up to about the 1890s, most of the land was owned and leased from the Crown. Large leasehold properties were taken up between 1850 and 1865. The holders of these leases were often called "runholders," and their properties were known as "runs" or "stations."

17. R. M. Burdon, *High Country: The Evolution of a New Zealand Sheep Station* (Auckland: Whitcombe and Tombs, 1938), 105.

18. E. C. Studholme, *Te Waimate: Early Station Life in New Zealand* (Wellington: Reed, 1940), 115–16.

19. Michael Mitton, diary, 30 June 1873, MB 44, box 4, Acland Family Papers, Macmillan Brown Library, University of Canterbury, Christchurch, New Zealand.

20. Acland Family Papers, box 51.

21. *Lyttelton Times*, 29 February 1860, 4.

22. *Lyttelton Times*, 17 May 1864, 4.

23. *Lyttelton Times*, 22 December 1863, 4.

24. Johannes C. Anderson, *Jubilee History of South Canterbury* (Auckland: Whitcombe and Tombs, 1916), 95.

25. Henry Ford Diary, 22 September 1872 and January 1873, the farm diaries of Pareora and Holme stations, October 1862–December 1873, Canterbury Museum Documentary Research Centre, Canterbury.

26. *Timaru Herald*, 26 July 1871, 2.

27. *Lyttelton Times*, 12 October 1861, 6.

28. *Lyttelton Times*, 25 October 1862, 5. A wether is a castrated ram. Ram lambs not required for future breeding purposes are castrated. Wethers are usually fattened for meat production but can also be kept for wool production since they grow more wool than ewes at the same level of feeding and can withstand harsher conditions.

29. *Timaru Herald*, 7 December 1870, 2.

30. *The Press*, 9 March 1872.

31. *The Press*, 29 October 1872.

32. *Lyttelton Times*, 13 May 1863, 3.

33. *Lyttelton Times*, 11 June 1862, 3.

34. *Lyttelton Times*, 20 May 1863, 5.

35. Oliver Scott Thompson, station diaries, April 1896, box 52, Acland Family Papers;; D. Livingston, stock tallies, 1911, box 62, Acland Family Papers.

36. *The Press,* 3 July 1875.

37. Noel Crawford, *The Station Years: A History of The Levels, Cannington, and Holme Station, with Special Reference to the Upper Regions of the Pareora River, Where They Joined* (Timaru: Noel Crawford, 1981), 36.

38. John McBeath, "Cross-Breeding of Sheep," *New Zealand Country Journal* 1, no. 4 (1877): 265–67.

39. "Correspondence," *New Zealand Country Journal* 1, no. 4 (1877): 274.

40. Davidson, *William Soltau Davidson, 1846–1924,* 23.

41. Ibid., 24–25.

42. Ibid., 25

43. Stevens, *James Little,* 7.

44. Ibid., 8–10

45. "Experiments by a Colonial Sheep-Breeder," *Livestock Journal,* reprinted in *New Zealand Country Journal* 7, no. 4 (1883): 302.

46. *Otago Witness,* 9 January 1907, 6.

47. Stevens, *James Little,* 12.

48. F. R. Marshall, "Features of the Sheep Industries of United States, New Zealand, and Australia Compared," *United States Department of Agriculture Bulletin No. 313* (Washington, DC: GPO, 1915), 21.

49. Ibid., 17–18.

50. Ibid., 1.

CHAPTER 13

Animal Science and the Representation of Local Breeds

Looking into the Sources of Current Characterization of Bororo Zebu

Saverio Krätli

GROWING INTERNATIONAL attention to the value of domestic animal biodiversity (DAD) has placed a strong emphasis on locally adapted breeds, particularly in developing countries.[1] The initiatives for cataloging DAD and prioritizing interventions for conservation have substantially relied on breed characterization—from the early Food and Agriculture Organization of the United Nations (FAO) inventories to the currently online DAD-IS database.[2] In fact, breed characterization not only informs the management of farm-animal genetic resources, but it is also deeply entrenched in rural-development policies and project design for the livestock sector. As characterizations define the productive value of local breeds, they also define the economic relevance of their producers. But what processes lead to the construction of scientific knowledge about locally adapted breeds?

This chapter embarks on a historical investigation of the sources of animal-science knowledge on the Bororo breed. It does so within the horizon of "practice" approaches in (1) political ecology—resources are defined by use patterns within networks of power;[3] and (2) sociological studies of science—knowledge is necessarily partial and located: mobilized in practices of social control, filtered by individual agency and technology, and constructed as universal through political strategies, by enrolling of actors and institutions in ever-wider knowledge networks.[4] With particular reference to Niger and the francophone tradition, the chapter looks at the scientific facts packed into the current characterizations of the Bororo breed, asking which *actor-network* produced them and through which processes.

The Bororo zebu kept by the WoDaaBe pastoralists is an emblematic example of a locally adapted breed in a low-input livestock system. Its extreme and unpredictable production environment and the specialization of its breeders are a guarantee of high biodiversity value.[5] Over the last forty years, the studies that have paid some attention to the Bororo have suggested that the scientific information on the breed need revising.[6] New findings departing from the received wisdom have never been incorporated into formal descriptions. The surprisingly persistent lack of reliable data on a breed with a population of several million across many countries has been repeatedly pointed out to no effect.[7] In Niger, extraordinary attention has been given to another local zebu, the Azawak. Initially bred only by a few Tuareg groups, the Azawak breed was increasingly taken over by expanding forces within the livestock sector—farmers and absentee owners—particularly following the major crisis of 1984.[8] Today, Niger's pastoral-development policy has the picture of an Azawak bull on the cover and is substantially geared toward this breed. The Bororo is mentioned once in the initial list of cattle breeds in the country and never again.[9] Yet, the Bororo represents a large proportion of the cattle population in the country and the breed most in demand on the export market.[10]

The marginal position of the Bororo within the development arena is explained by specialists and administrators on scientific bases: lowest rank in milk production, meat quality, and fertility rate; poor dressing percentage; a semi-wild nature that makes the breed difficult to handle and useless for work. Reference is often made to "several studies" that would have "proved beyond doubt" the inferior performance of the Bororo and, consequently, its negligible economic value. None of the specialists and administrators I interviewed, however, was able to identify such studies with any precision. How were these studies carried out, under which conditions, and how robust are their findings?

Received Wisdom

The problem of identifying primary sources of data exists above all for the Bororo, as the official data on Azawak consistently refer to the herds selected since 1933 at the research station of Filingué/Toukounous.[11] Since the first study by Jean Pagot, the Azawak has been the topic of dissertations for generations of students from Niger (in graduate courses in veterinary medicine and agronomy and at the École des Cadres de l'Élevage, the technical school of the livestock service outside Niamey).[12] These works are usually presentations of a station-based measurement exercise, interpreted in light of previous data and framed in a general description of pastoral systems and cattle breeds in Niger. They usually include a description of the Bororo. As students become consultants and/or take up managerial posts with the government or with international projects, this body of data makes its way into more authoritative sources. Over the years, repetition and the habit of chain referencing have generated the impression of a substantial amount of knowledge available not only on the Azawak but also on the Bororo. However, a careful inspection of the chain of references leads to only one source: George Doutressoulle's 1947 very comprehensive work on the livestock sector in French West Africa (*L'élevage en Afrique Occidentale Française*, henceforth *L'élevage en A.O.F.*). As stated by the author himself (in the foreword), this is a reference work, not a primary source. Nowadays, *L'élevage en A.O.F.* is relatively difficult to access. Descriptions of Bororo and Azawak are available in two other works that are commonly found in the offices of the Ministry of Animal Resources (MAR) in Niger, even outside Niamey. These are the official pocket handbook for agronomists (the *Mémento de l'agronome*, 1980, henceforth *Mémento*)[13] and René Larrat's veterinary field manual.[14] Both works are published by the French government's development agency (Ministère de la Cooperation) and subsidized for francophone developing countries. Their descriptions of the Bororo and the Azawak include exact production figures such as daily milk average and dressing percentage, as well as more general evaluations. These data consistently back up the information I gathered from my interviews with MAR personnel.[15] Although neither of these works provides the sources of the data presented on the Bororo, a comparison side by side with Doutressoulle's text reveals substantial word for word overlapping, leaving no doubt about their debt to the work of 1947.

Fortunately, *L'élevage en A.O.F.* includes a list of bibliographical sources against which its description of the Bororo can finally be checked. Five works from this list (out of twenty-four) contain information on the Bororo. They are all by French veterinarians and date from 1906 to 1941. Three of them (Pécaud, Malbrant, and Mornet and Koné) include milk-production figures.[16] Of the remaining two studies, one is Doutressoulle's final dissertation in veterinary medicine, an overview of the livestock sector in Niger.[17] The other is the early overview of animal husbandry in West Africa by "Vétérinaire-Colonel" Pierre, the work that first introduced the notion that Bororo are poor producers: "The meat is hard and stringy. The females . . . hardly feed their calves."[18] Problems concerning the precision of Pierre's description of the Bororo (under the name of "Fogha zebu")[19] were promptly outlined by George Pécaud.[20] Later on, Paul Mornet and Kassoum Koné argued that Pierre had mistakenly based his description on the picture of a young animal.[21] The descriptions of the Bororo given by George Pécaud and Réné Malbrant are significantly more positive.[22] Referring respectively to animals in Benin and Chad, these authors quoted a milk production of six to seven liters per day during the good season.[23] On the other hand, the young Doutressoulle subscribed to Pierre's negative judgment, going even further in dismissing all transhumant pastoralism as inherently unproductive.[24]

Mornet and Koné published their article on the Bororo zebu in the bulletin of the AOF livestock service in 1941.[25] To the best of my knowledge, this is the first and only scientific study ever dedicated to the breed. Both men were veterinarians. Mornet, who had been interim chief of the Niger livestock service in 1937, had recently been transferred to Dakar. Koné, one of the first African veterinarians in the AOF, had worked for several years at the abattoir of Niamey before being appointed chief of the outpost of N'Guigmi, on the border with Chad. In that crucial transhumance point, Koné worked in close contact with Bororo herds. Their study introduced some remarkable insights. The Bororo is described as perfectly adapted to the sahelian environment and to the mobility of their herders who "in order to keep their zebu cattle in good condition . . . continue this endless wandering and, despite the unrewarding nature, every year match the challenge of keeping alive animals whose nutritional needs (given their size) exceed the potential of the range."[26] The authors mention the Bororo's exceptionally large and thick hide and address the issue of the breed's value as a beef animal with reference not to metropolitan taste but to its regional market: "Bororo herds supply part of Nigeria's markets, where they sell at a premium. The indigenous butcher, like the consumer . . . what they like is the important muscular mass of this zebu, whose weight is by far higher than that of the others."[27]

The study is equally sharp on the issue of milk production, for the first time expressed with reference to the season: the Bororo's rainy season average is recorded as five-to-six liters per day, dropping to two-to-three liters per day in the dry season.[28] These figures can be understood in perspective if compared to the Azawak's daily production record at the breeding station during the exceptionally good year of 1938: 7.5 liters.[29] Despite the unique content of the study by Mornet and Koné and the institutional context of its publication, these new data on the Bororo were never taken into consideration, not even for criticism.[30]

In *L'élevage en l'A.O.F.*, Doutressoulle did not use the data on the Bororo from most of the literature he referred to in his bibliography, nor did he discuss them. Instead, he maintained his own early view that the breed was aesthetically impressive but had no productive value. He supported this view with precise figures on production, but for which he provided no source. On milk yield, he wrote that "the average lactation period is six months and varies from 3 to 4 liters in the best animals at the beginning of lactation, and drops to 1.5 liters at the end."[31] He also stuck to Pierre's ill-informed opinion that the Bororo's potential as a beef animal must be low: "poor potential as a beef animal due to the relative size of its skeleton."[32] While failing the Bororo, *L'élevage en l'A.O.F.* promoted the Azawak to "the highest rank among our dairy cows,"[33] a formula paraphrased in almost every work on the breed ever since, as well as being reiterated in each edition of the *Mémento* until that of 2002.[34]

The Azawak had been bred at the research station of Filingué since 1931, but the first systematic study on production had been carried out, by Jean Pagot, in the early 1940s. Working with cattle that had been allowed as calves to eat at will and enjoy an "improved" feeding regime, Pagot had found an average daily milk production of 1.6 liters, with a minimum of 0.8 liters per day in January and a maximum of 2.8 in August.[35] Although Doutressoulle was aware of milk production measurements at Filingué, the description of the Azawak in *L'élevage en l'A.O.F.* stated that "its lactation period is of seven to eight months on average and reaches 800-to-1,000 liters."[36] These figures are discussed here only for their comparative value, in order to show how the data in this crucially influential work did not correspond to the given sources. Instead, new figures were introduced (for which no reference was given) contrary to existing data and presenting the Azawak as a superior breed and the Bororo as showy but worthless.

The Wider Setting

L'Élevage en A.O.F. is more relevant to our concern for its long-standing influence and for what it says about the process of construction of scientific breed characterization than for the way it appears to have filtered the information at the time of its publication.[37] Insofar as *L'élevage en A.O.F.* spun the data in favor of the Azawak, this only conformed to an entrenched attitude on the part of the administration that, in Niger, had begun well before 1947.

According to the archival records, a breeding program for the improvement of the Azawak and the dissemination of selected bulls "in the traditional system" was a key objective of the Filingué center since its creation by the agricultural service as "Sahelian Fodder Experimental Station" in 1931. When the station was handed over to the livestock service in 1937, the service inherited the breeding program and a herd of about ninety head.[38] But on which basis was the Azawak chosen in 1931? Here are the descriptions of Azawak and Bororo as given in the report of the agricultural service for 1931:

> A/ The zebu Azaouak—medium size—1.25 to 1.30 m—the hump and the dewlap are little developed, the head is quite short, and the horns are thin, thin skin and members, the coat is usually pale, darker in bulls, temperate and hardy; average weight from 500 to 650 kg, the zebu Azawak are usually in the hands of the Daoussaks or the Bellas.[39]
>
> B/ The zebu Borodji [*sic*]—of larger size than the previous one, hump and dewlap very developed, strong head and big horns, thick skin and heavy skeleton; coat is dark red, less resistant than the previous one. The Borodji, are almost exclusively owned by the FulBe.[40]

The statement that Bororo are less resistant than Azawak is striking, as resistance was the only virtue of the Bororo that even its detractors had been willing to acknowledge. In Pierre's words, "Accustomed to harsh weather and to alternate conditions of abundance and shortage, this breed is necessarily robust and hardy."[41] Even more puzzling, the average weight attributed to the Azawak is twice as high as in all later sources[42] and more suggestive of the larger Bororo.[43] This mismatch concerning the weight is important, as it was on this feature and not on milk production that the report recommended the Azawak for the breeding program: "The Azaouak zebu is by far the most interesting of these three types—it is fit for beef production but dairy traits are poorly pronounced: the udder is little and show signs of poor blood supply."[44]

Let us, therefore, summarize the scientific facts behind the crucial decision to choose the Azawak for the breeding program in Filingué in 1931 (starting a snowball effect of interest within animal science and international development).[45] The breed that was soon to be at the core of the livestock service program of selection for milk production (started in 1937 and still going on) was actually described as having poor dairy potential in 1931. The decision to introduce the Azawak in Filingué was taken following a report that recommended it as a beef animal, but on the basis of dramatically incorrect weight figures. At the same time, the larger Bororo, preferred on the regional meat market, was excluded as uninteresting.[46] It is worth noting how the report, although offering a rather meager characterization of the breeds, found room to specify which ethnic group kept them. Was this information meant to back up the decision?

Animal Science with a Human Heart

In 1931, the administration had hardly any knowledge of the Azawak. In fact, while the Bororo breed had been repeatedly described as unusually homogeneous,[47] as late as 1943 there were cautions on the actual existence of an "Azawak breed."[48] The report recommending them as "by far the most interesting breed" is probably the first official document to use the toponym "Azaouak" with reference to a type of cattle.[49] Before then, there had merely been confused remarks. One can only conjecture, for example, that the animals that later became known as Azawak were behind a reference to the "race d'Azbin" in the 1926 report of the livestock service by "vétérinaire de 3e classe M. Diard" (the only European officer in a livestock service with a staff of eight across all Niger):

> In the Adrar [Ader], district of Tahoua, sedentary people and Tuaregs have crossbred these two breeds [Goudali and Bororo] obtaining a nice product with higher yields in milk and meat. The Degamenas tribe ("maraboutique" Tuaregs) has taken this route and owns the best herds in the country. The Igdaleuss breed a small cow (the Azbin breed) that gives a good product and is quite a good milker.[50]

Extracted word for word from Doutressoulle's dissertation, this favorable description seemed to have been initially quite controversial within the Niger livestock service itself.[51] When Pécaud became director, the following year, he made a point of rectifying it according to his own view:

> The Arab zebu is found above all in the countries East of Gouré or in the North. The "small cattle of the Azbin," the animals called "aznadji" . . . bred in the north by Tuareg or arabized people, relate back to this breed, more or less modified, . . . weak dairy potential: three to four liters per day, . . . mediocre fattening capacity. It is a very poorly fixed breed, whose local varieties are very numerous."[52]

The "Dégamenas" mentioned by Diard/Doutressoulle were most likely the Dar'Menna (also spelled Daghmenna) of the Ineslemeden. The Dar'Menna were one of the clans that collaborated with the French during the 1916–17 uprising of the Tuareg aristocracy (*imajeghen*), started in the sector of Menaka (today in Mali) and spread eastward along the Azaouak Valley and into the Aïr.[53] "Azbin" was another name for today's Aïr, where new sedentary groups of Tuaregs resulting from the political adjustments that had followed the defeat of the *imajeghen* were perceived by the administration as examples of collaboration:

> The sedentary Tuaregs have settled in the valleys of the Aïr.... They have almost acquired the love for the soil and the mentality of our French farmers. They appear to be hard workers and they like money. As soon as they develop some commercial entrepreneurship, which at the moment they utterly lack, they will be true *paysans* [small holders].[54]

The colonial administration took the view that the promotion of sedentary agriculture in the pastoral zone was associated with political "pacification." It seems hardly a coincidence that the first agricultural research station in Niger, with a strong "educational" mission, was placed in 1931 right at the epicenter of the 1916–17 uprising.[55] Most likely, the need to respond to the delicate political situation of the years following the insurgency affected decision making in the matter of pastoral development. As the colonial veterinarians were also in charge of development policies and implementation, they must have found it natural to legitimate in animal-science terms their political necessities.[56]

This, however, need not be the only explanation for the preference given to the Azawak, which should be understood in the context of contemporary pastoral-development theories. The station of Filingué was "created with the general goal of prompting the rational development of livestock-breeding in the sahelian region." The main way forward in that direction was seen in fodder improvement, but one of the strategies included "the at least partial suppression of nomadism, and later on, of transhumance."[57] This orientation would have been enough automatically to disqualify the Bororo in the eyes of the administrators as the breed selected by the most mobile livestock system in the country, independent of any consideration of its performance.

The core argument of French pastoral-development policy in West Africa went as follows: increasing production depends on improving feeding conditions; in turn, this can only be achieved through agricultural methods, namely, fodder cultivation and the adoption of feed integrators of agricultural origin. As summarized in the words of the General Secretary of Haut-Sénégal Niger Jacques Méniaud, "We should not rely too much on crossbreeding trials with superior breeds and selection of breeding bulls if, at the same time, we do not improve the feeding conditions of the herds, and this is a matter of agricultural production."[58] This view was deeply entrenched in animal-science disciplinary commitment to a program of agricultural intensification.[59] Although there were notable exceptions to this rule,[60] the overarching policy orientation prescribed the replacement of transhumant, specialized pastoralism with "crop-livestock integration."[61] This matched closely the model of livestock breeding in France at the beginning of twentieth century.[62] It was, therefore, only natural for the administrators to see this process as the way to "rationalize" the livestock sector in the colonies. Doutressoulle echoed this approach in his early study of the livestock sector in Niger:

> With perseverance we should obtain some result among the sedentary populations, introducing, for livestock nutrition, the use of agricultural by-products such as stalks of niebe beans and of groundnuts, and potato leaves. . . . But concerning the nomads, the Peuhls [WoDaaBe], and they own the largest proportion of herds, we should not dream of fodder reserves: lack of labor, lack of stability, herds that are too strong. The sedentarization of herders and their transformation into agro-pastoralists is the only way to compensate for soil degradation, integrating livestock nutrition with the produce of intensive cultures.[63]

Carriers of cultural sensitivity and administrative necessity, informed by contingent constraints, these scientific studies characterizing the Bororo and the Azawak offered explicit value judgments framed in anthropomorphic imagery, together with morphological descriptions, body measurements and production figures. The Bororo were described as feral and unmanageable animals, "semi-wild . . . the udder . . . covered by coarse hair"[64] and as a "primitive breed . . . one of the most unrefined breeds . . . nomadic par excellence."[65] On the other hand, the Azawak were said to love town and described as sedentary cattle. A young French veterinarian, reporting on information collected during his work at Toukounous in the 1960s, wrote of the Azawak that "their sedentary nature, contrary to the migratory and restive nature of the Bororo zebu, make these animals the best breed for both the farmer and the herder in the process of settling."[66] Crossbreeding with a Bororo, on the other hand, would give an Azawak "a more independent attitude."[67] In fact, observation would have shown the opposite: in traditional herding systems, Azawak animals go for days unattended, while Bororo are literally addicted to the constant presence of the herder.[68] The idea that the Bororo have an "independent" nature probably resulted from a habit to associate the breed with wildlife and, more generally, to identify the animals with the image of their herders within the sedentary groups that interfaced with the administration. In the late 1930s, Kassoum Koné included in one of his reports a few pages of information about the public image of the WoDaaBe. Here is a long, enlightening extract:

> Among the Peuhls [FulBe] of Niger, the Peuhls Borrorodji [WoDaaBe] constitute a subgroup with a bad reputation from the social, religious, and moral point of view. The other local races look at them as inferior beings, in the same way as the animals carrying their name: the borrorodji [*sic*] cattle. They [the WoDaaBe] are relegated to the rank of wild animals who flee as man approaches. They are considered lawless beings, without tradition and without home, unsociable, carrying out an eternally nomadic existence across the countries, having no other horizon than the thick bush and no other companion than their cattle. They are considered as scavengers who feed on rotting corpses of animals that died of natural causes. All evil charms and extraordinary powers are ascribed to them: they are hyena-men, capable at night of transforming themselves into hyenas in order to attack other people's herds out of greed and envy. They are considered godless, with no religion at all, not even animism, just like animals. Therefore, they are banned from human society. They have all the vices and all the faults, not a single virtue. They are considered as bastards, out of illegitimate marriages, as there is no marriage among them but rather half-marriages, unstable, left to women's whims. Women and girls chose their men in a line of youths and the choice is driven by beauty. There is no dowry. The bride may leave her partner overnight, for another one more handsome. Adultery goes unpunished.[69]

Apart from rare exceptions—the most notable of which resulted in the study by Mornet and Koné—French veterinarians had very little chance to see Bororo herds from up close (something quite difficult even today without the active cooperation of the herder).[70] French veterinarians were extremely few in number and lacked mobility within an immense and harsh environment.[71] For information on all that was beyond their reach, they were totally dependent upon their interpreters and local staff who, most likely, neither liked the herders of the Bororo nor knew much about them. Did the popular image of the WoDaaBe as wild people—therefore, "irrational"—continue, even in subsequent years and despite the available evidence, to reflect on the perception and representation of their economic life, of which their cattle are, of course, both the productive core and the most prominent symbol?[72]

OUR SEARCH for the sources of animal-science knowledge on the Bororo zebu has shown that it dates as far back as the early twentieth century, rooted in confused information and misrepresentations, all to the effect of discrediting the breed in comparison with the Azawak. Several forces appear to have contributed to steer scientific judgment in this direction: a prejudicial misunderstanding of specialized pastoralism based on the French model of cattle keeping as a practice merely marginal and ancillary to agriculture; the interest on the part of the administration to strengthen their links with the few settled and loyal Tuareg tribes in a region still affected by great political instability; a prejudicial view of nomadism in general and of the WoDaaBe in particular (reflecting on the administration's perception of the herders' economic strategies). A negative image of the nomads was embedded in the administration's perspective as well as locally induced, both through daily exposure to the cultural mediation of people from sedentary ethnic groups and by the WoDaaBe's reserved attitude and common strategy to keep a low profile, particularly in the early years of their migrations into Niger.

If the most specialized groups of pastoralists could not be settled, perhaps livestock could. So, efforts were made to expand the sedentary cattle-keeping sector, which was characteristically easier to reach, control, and influence. The veterinarians' conceptual appropriation in the 1920s of a little-known variety of sahelian zebu kept by (nomadic) Tuaregs and its reinvention as the "sedentary" Azawak breed became the instrument to pursue such a goal.

Notes

This chapter is based on research conducted for my PhD dissertation on the WoDaaBe's cattle-breed selection in Niger: S. Krätli, "Cows Who Choose Domestication: Generation and Management of Domestic Animal Diversity by WoDaaBe Pastoralists (Niger)" (PhD diss., Institute of Development Studies, University of Sussex, Brighton, 2007).

1. UN, *Convention on Biological Diversity* (Rio de Janeiro: United Nations, 1992); Food and Agriculture Organization of the United Nations, *The Global Strategy for the Management of Farm Animal Genetic Resources: Executive Brief* (Rome: FAO, 1999); FAO, *The State of the World's Animal Genetic Resources for Food and Agriculture*, ed. Barbara Rischkowsky and Daffyd Pilling (Rome: FAO, 2007). See also S. J. G. Hall, *Livestock Biodiversity: Genetic Resources for the Farming of the Future* (Oxford: Blackwell, 2004).

2. N. R. Joshi, E. A. McLaughlin, R. W. Phillips, *Types and Breeds of African Cattle*, FAO Agricultural Studies no. 37 (Washington: FAO, 1957); FAO DAD-IS, Domestic Animals Diversity Information System, http://dad.fao.org (accessed 2 April 2009).

3. M. Leach and R. Mearns, *The Lie of the Land: Challenging Received Wisdom on the African Environment* (Oxford: James Currey, 1996); A. E. Nyerges, "Introduction—The Ecology of Practice," in *The Ecology of Practice: Studies of Food Crop Production in Sub-Saharan West Africa*, ed. A. E. Nyerges (Amsterdam: Gordon and Breach, 1997), 1–38.

4. B. Latour, *Science in Action* (Cambridge, MA: Harvard University Press, 1987); idem, "Le métier du chercheur: Regard d'un anthropologue" (paper presented at a conference held by Institut National de la Recherche Agronomique [INRA], Paris, 22 September 1994); J. Fairhead and M. Leach, *Science, Society and Power: Environmental Knowledge and Policy in West Africa and the Caribbean* (Cambridge: Cambridge University Press, 2003).

5. It is within unpredictable environments that the highest levels of within-breed variation have been observed. D. E. MacHugh, M. D. Shriver, R. T. Loftus, P. Cunningham, and D. G. Bradley, "Microsatellite DNA Variation and the Evolution, Domestication and Phylogeography of Taurine and Zebu Cattle (*Bos taurus* and *Bos indicus*)," *Genetics* 146, no. 3 (July 1997): 1071–86; E. M. Ibeagha-Awemu and G. Erhard, "An Evaluation of Genetic Diversity Indices of the Red Bororo and White Fulani Cattle Breeds with Different Molecular Markers and Their Implications for Current and Future Improvement Options," *Tropical Animal Health and Production* 38, no. 5 (2006): 431–41.

6. Two papers presented at the conference "Colloque sur l'élevage," Fort-Lamy, Chad, 8–13 December 1969: P. Capitaine, "Projet d'amélioration de zébu M'Bororo par croisement avec le zébu Foulbe de N'Gaoundere"; and P. Lhoste, "Les races bovines de l'adamaoua (Cameroun)". See also, G. Tacher, "Les abattages de bovins à l'abattoir frigorifique de Farcha (Fort-Lamy) de 1967 à 1970: Analyse statistique et interpretation," in *Région de recherches vétérinaires et zootechniques d'Afrique Céntrale: Rapport annuel 1971* (Chad: Laboratoire

de Recherches Vétérinaires de Farcha, 1971), 267; A. Bonfiglioli Maliki, *Ngay-naaka: Herding According to the WoDaaBe*, Discussion Paper no. 2, Republic of Niger, Ministry of Rural Development, Niger Range and Livestock Project (Tahoua: USAID/Niger, 1981); and J. J. Delate, H. Ouyan, and S. Theander, "Influence de l'âge, du sexe, de la race sur l'embouche des zébus nourris avec des sous-produits rizicoles dans le Nord Cameroun," *Revue d'Élevage et de Médecine Vétérinaire des Pays Tropicaux* 39, no. 1 (1986): 89–95.

7. S. L. Louis, *Technical Report on Animal Production in the Pastoral Zone of Niger: Recommendations for Improvement* (Tahoua, Niger: Niger Range and Livestock Project, 1982), 62; D. Bourn, W. Wint, R. Blench, and E. Wolley, *Nigerian Livestock Resources*. vol. 2, *National Synthesis* (St. Helier, Jersey, U.K.: Resources Inventory and Management Limited, 1992), 85.

8. J. Nicolaisen, *Ecology and Culture of the Pastoral Tuareg: With Particular Reference to the Tuareg of Ahaggar and Ayr* (Copenhagen: National Museum of Copenhagen, 1963), 50; A. Habou and A. Danguioua, "Transfert du capital-bétail au Niger des pasteurs aux autres catégories socio-professionnelles: Illusion ou réalité. Rapport Préliminaire de Mission (21.12.90/26–01.91)" (University of Wisconsin Land Tenure Center—Secretariat Permanent du Comité National du Code Rural, 1991).

9. *Document cadre pour la relance du secteur élevage au Niger: État des lieux, axes d'intervention et programmes prioritaires, novembre 2001* (Niamey: Ministère des Resources Animales, 2001).

10. B. Djariri and M. Saley, with H. B. Dahiru, *L'adaptation des circuits de commercialisation des bovins nigériens à l'évolution de la demande au Nigeria: Suivi des échanges transfrontaliers entre le Nigeria et les pays voisins* (Niamey: LARES and IRAM, 2003).

11. The station was created in 1931 in Filingué (about 200 km north of Niamey), then moved in 1954 some 18 km north, to Toukounous.

12. J. Pagot, "Le zébu de l'Azawak," *Bulletin des Services Zootechniques et des Épizooties de l'A.O.F.* 6, nos. 1–4 (1943): 155–63; idem, "Production laitière en zone tropicale: Faits d'expérience en A.O.F.," *Revue d'Élevage et de Médecine Vétérinaire des Pays Tropicaux*, n.s., 5, special issue (1952): 173–90.

13. Various authors, *Mémento de l'agronome*, 3rd ed. (Paris: Ministère de la cooperation, 1980).

14. R. Larrat, *Manuel vétérinaire des agents techniques de l'élevage tropical* (Paris: SEAE, 1971).

15. The same data inform the Niger record on the Bororo within DAD-IS, the database developed by FAO as part of its global strategy for the conservation of farm-animal genetic resources. FAO, DAD-IS—Domestic Animals Diversity Information System, http://dad.fao.org/ (accessed 2 April 2009).

16. G. Pécaud, *L'élevage et les animaux domestiques du Dahomey* (Dakar: Imprimerie du Gouvernement Général A.O.F., 1912); R. Malbrant, "La production animale au Tchad" (report presented at the Congrès de la Production Animale et des Maladies du Bétail, Paris, 17–18 June 1931); P. Mornet and K. Koné, "Le zébu peulh bororo," *Bulletin des Services Zootechniques et des Épizooties de l'A.O.F.* 4, nos. 3–4 (1941): 167–80.

17. G. Doutressoulle, "L'élevage au Niger (A.O.F.)" (PhD diss., Université de Toulouse, 1924).

18. C. Pierre, *L'élevage dans l'Afrique Occidentale Française* (Paris: Gouvernement Général de l'Afrique Occidentale Française, 1906), 106.

19. The Dallol Fogha (Fogha Valley), on the northern bank of the Niger River, about 25 km north of Gaya. Pierre (ibid.) described the breed as native to that region.

20. Pécaud, *L'élevage et les animaux domestiques,* 43.

21. Mornet and Koné, "Le zébu peulh bororo," 171.

22. Pécaud, *L'élevage et les animaux domestiques,* 45; Malbrant, *La production animale au Tchad,* 64.

23. When, later on, Pécaud was given the direction of the Niger livestock service, he confirmed these figures; Pécaud, *Rapport du chef du service zootechnique de la colonie du Niger sur l'état sanitaire des animaux et de le fonctionnement du service pendant l'année 1927* (Niamey: Service Zootechnique et des Épizooties, 1928).

24. Doutressoulle, "L'élevage au Niger," 39, 42, 107–8.

25. Mornet and Koné, "Le zébu peulh bororo," 171.

26. Ibid., 176. All translations are by the author.

27. Ibid., 179.

28. Ibid., 179–80.

29. "Qualités laitières de la race bovine de l'Azaouack" *Bulletin des Services Zootechniques et des Épizooties de l'Afrique Occidentale Française* 1, no. 2 (1938): 51. A study of the Azawak in traditional settings a few years later, argued that the recorded milk yield of two to three liters per day in the good season was "a very acceptable average in such harsh conditions." A. Couture, "Contribution à l'ethnologie du zébu dit 'de l'Azawak,'" *Bulletin des Services de l'Élevage et des Industries Animales de l'Afrique Occidentale Française* 1, no.1 (1948): 46.

30. The bibliography in *L'élevage en A.O.F.* misquotes it as "Mornet et Kassoun Koni, *Le zèbre peul Bororo.*"

31. G. Doutressoulle, *L'élevage en Afrique Occidentale Française* (Paris: Larose, 1947), 106.

32. Ibid., 106.

33. Ibid., 104.

34. The *Mémento de l'agronome* still maintained the same descriptions in the edition of 1991. The latest edition (2002) has changed format, eliminating the entire section on breeds.

35. Pagot, "Le zébu de l'Azawak," 158. More measurements carried out at Filingué were published a little after Doutressoulle's work appeared in print. Average production of a sample of "good milkers" was 2.6 liters per day, while other groups' performance was 1.5 to 1.7 liters per day; BSEIA, "Le contrôle laitièr des zébus Azawak à la station d'élevage de Filingué (Niger)," *Bulletin des Services de l'Élevage et des Industries Animales de l'Afrique Occidentale Française* 1, nos. 2–3 (1947): 91–94. A 1950 report by Pagot, analyzing Azawak milk production at Filingué over a seven-year span, gives an average daily production of 1.88 liters; J. Pagot, *Rapport à OSTROM n. 71 GA—4.11.1950* (Service de l'Élevage et des Industries Animales de l'Afrique Occidentale Française, Office de la Recherche Scientifique Coloniale, Ségou, Mali, 1950).

36. Doutressoulle, *L'élevage en Afrique Occidentale Française,* 104.

37. One should not lose sight of the fact that the descriptions of the Bororo and the Azawak in *L'élevage en A.O.F.,* although magnified here by the focus of the analysis, occupy only a couple of pages out of three hundred in a work covering six species, from cattle to chicken and ducks. Doutressoulle was unlikely to have any interest in promoting one breed over the other, apart perhaps from an attachment to his own first impression when he was working in Niger in the early 1920s.

38. Service de l'Agriculture, *Rapport annuel agricole 1931* (Niamey: Service de l'Agriculture, 1932); Mornet, *Rapport annuel 1937* (Niamey: Service Zootechnique et des Épizooties, 1938).

39. The Daoussaks (also Dawsaak, Dahusahak, Idaksahak) are found between Menaka (Mali) and Tillaberi/Filingué, integrated among the Tuareg Imghad (but originally in a position of vassalage) and linguistically midway between Tamasheq and Songhai-Zerma. The term "Bella" (Songhai), like "Buzu" (Hawsa), designated servants (captives) in Tuareg society. E. Bernus, *Touaregs nigériens: Unités culturelles et diversités régionales d'un peuple pasteur* (Paris: O.R.S.T.O.M., 1981), 393 and 62.

40. Today more commonly "Peuls," the French word for FulBe (from the singular Pullo), the wider Fulfulde-speaking group including the WoDaaBe. Service de l'Agriculture, *Rapport annuel agricole 1931,* 14.

41. Pierre, *L'élevage dans l'Afrique Occidentale Française,* 107.

42. For selected animals grown at Filingué: 300–400 kg, Pagot, "Le zébu de l'Azawak," 157. For a general average: 250–350 kg, R. Larrat, *Problèmes de la viande en A.O.F. zones de production: II—Niger* (Paris: Èditions Diloutremer, 1955), 11.

43. For animals from the bush: average 450 kg, maximum 700 kg, according to Mornet and Koné, "Le zébu peulh bororo," 171; maximum 450–550 kg, according to Larrat, *Problèmes de la viande en A.O.F.*, 11.

44. Service de l'Agriculture, *Rapport annuel agricole 1931*, 17.

45. Recently culminated in the Azawak Project, which reinvented the 1931–37 program of dissemination of selected bulls within the "traditional" system; *Projet Azawak, Appui à l'élevage des bovins de race Azawak au Niger: Rapport de formulation* (Bruxelles: Cooperation Technique Belge et Vétérinaires sans Frontières, 2001).

46. Although this position was soon to become dominant, not everybody within the administration shared it in the same measure in those early years. The director of the Agricultural Service in 1934, for example, was of the opinion that the presumed lack of economic value of the Bororo needed some further investigation and made arrangements for keeping a small herd under observation at Filingué. According to the annual report for 1935, "No doubt, if sufficiently cared for, nourished and watered, these animals will very quickly adjust to their new existence"; R. Soulat, *Rapport agricole annuel 1935* (Niamey: Gouvernement Général de l'Afrique Occidentale Française, Colonie du Niger, Service de l'Agriculture, 1936), 37. In 1936, the center of Filingué was passed over to the livestock service, and in the following years, we find no further mention of this project.

47. For example, Pécaud, *L'élevage et les animaux domestiques*. Mornet and Koné describe the Bororo as "a perfectly characterized breed"; Mornet and Koné, "Le zébu peulh bororo," 170.

48. "As far as the Azawak zebu is concerned, the expression 'breeds of the Azawak region' is to be preferred to 'Azawak breed' . . . Without taking into consideration Bororo crossbred, we find [within the Azawak breed] animals with very similar morphological traits but huge phaneroptical differences"; Pagot, "Le zébu de l'Azawak," 155. Phaneroptical traits refer to the colors of coat and skin.

49. According to Mornet, "The name 'Azawak' was given by the veterinarians to the kind of Arab zebu found in the regions of Menaka, Filingué, Tahoua, etc." P. Mornet, "Amélioration de l'élevage: Rôle des établissements d'élevage: Extension de l'élevage en zone sud," *Bulletin des Services de l'Élevage et des Industries Animales de l'A.O.F.* 4, no.1 (1951): 130.

50. M. Diard, *Rapport annuel du Service Zootechnique et des Épizooties du Niger 1926* (Niamey: Gouvernement Centrale de l'Afrique Occidentale Française, Colonie du Niger, Service Zootechnique et des Épizooties, 1927), 2.

51. Doutressoulle, "L'élevage au Niger," 42.

52. Pécaud, *Rapport du chef du Service Zootechnique*, 5.

53. F. Nicolas, *Tamesna: Les Loullemenden de l'est ou Touareg "Kel Dinnik."* *Cercle de Tahoua, Niger (Linguistique et ethnologie berbère)* (Paris: Imprimerie National, 1950); R. Alfarouk, "La politique coloniale d'affaiblissement de la confédération Kel Denneg de 1900 à 1949," in *Nomades et commandants: Administration et sociétés nomades dans l'ancienne A.O.F.,* ed. E. Bernus (Paris: Harmattan, 1993), 87–92; Bernus, *Touaregs nigériens*; idem, "Nobles et religieux: L'intervention coloniale dans une rivalité ancienne (Iwellemmedan Kel Denneg)," in *Nomades et commandants,* 61–68.

54. Col. M. Abadie, *La colonie du Niger* (Paris: Société d'Éditions Géographiques, Maritimes et Coloniales, 1927), 262.

55. Filingué, at the time of the rebellion an administrative outpost in the sector of Ménaka, had been the theater of the first major fighting. Nicolas, *Tamesna.*

56. E. Landais, "Sur les doctrines des vétérinaires coloniaux français en Afrique noire," *Cahiers des Sciences Humaines* 26, nos. 1–2 (1990): 33–71.

57. Service de l'Agriculture, *Rapport annuel agricole 1931,* 14.

58. Quoted in L. Delpy, "Le chemin de fer transsaharien et l'élevage en Afrique Occidentale," *Recueil de Médecine Vétérinaire Exotique* 6, no. 2 (1933): 153.

59. On the theoretical framework of early animal science, see B. Denis and M. Théret, "Les grands traités de zootechnie et leur conception de cette discipline," *Ethnozootechnie* 54 (1994): 3–24; R. Jussiau and L. Montméas, "La zootechnie: Une discipline d'enseignement vue à travers les manuels scolaires," *Ethnozootechnie* 54 (1994): 57–75; and E. Landais and J. Bonnemarie, "La zootechnie: Art ou science? Entre nature et société, l'histoire exemplaire d'une discipline finalisée," *Le Courrier de l'Environnement* 27, no. 4 (1996): 23–44.

60. Delpy, for example, recommended a "different formula: supporting the extensive livestock system practiced by the nomads. . . . Only such a system can be remunerative, as its distance from the commercial outlets results in significantly lower sale prices for meat, skins, and wool." "Le chemin de fer transsaharien," 154–55.

61. M. Piettre, "Les bases d'un grand élevage colonial (1)," *Recueil de Médecine Vétérinaire Exotique* 3, no. 3 (1930): 125–37.

62. J. L. Mayaud, "L'élevage bovin: D'un mal nécessaire à la spécialisation" in *Le mangeur et l'animal: Mutations de l'élevage et de la consummation,* ed. M. Paillat, Coll. Mutations/Mangeurs, no. 172 (Paris: Autrement, 1997), 11–32.

63. Doutressoulle, "L'élevage au Niger," 69.

64. Pierre, *L'élevage dans l'Afrique Occidentale Française,* 105.

65. Doutressoulle, *L'élevage en Afrique Occidentale Française,* 183.

66. J. L. Simoulin, "Le zébu de l'Azaouak: L'amélioration de l'élevage en zone sahélienne" (PhD diss., Faculté de Médecine et de Pharmacie de Lyon, 1965), 109.

67. Ibid, 62.

68. More details on this characteristic of the WoDaaBe cattle breeding and management system can be found in S. Krätli, "Cattle Breeding, Complexity and Mobility in a Structurally Unpredictable Environment: The WoDaaBe Herders of Niger," *Nomadic Peoples* 12, no. 1 (2008): 11–41.

69. Quoted in B. Hama, *Contribution à la connaissance de l'histoire des peul* (Niamey: Publications de la République du Niger, 1968), 53.

70. Mornet and Koné, "Le zébu peulh bororo."

71. There was only one veterinarian in Niger in 1927, based in Niamey, and forty-nine in the entire AOF twenty years later.

72. For a brilliant overview of the deeply rooted negative representation of the "nomad" in the European tradition, see P. Trousset, "L'image du nomade saharien dans l'historiographie antique," *Production Pastorale et Société* 10 (1982): 97–105. More specifically about the French colonial administration in (North) Africa and a view of nomads predominantly within a "civilization–versus-barbarism" dichotomy, see D. K. Davis, "Desert 'Wastes' of the Maghreb: Desertification Narratives in French Colonial Environmental History in North Africa," *Cultural Geographies* 11, no. 4 (2004): 359–87; and Diana K. Davis, *Resurrecting the Granary of Rome: Environmental History and French Colonial Expansion in North Africa* (Athens: Ohio University Press, 2007).

CHAPTER 14

Kenya's Cattle Trade and the Economics of Empire, 1918–48

David Anderson

THE STUDY of the history of livestock production in Africa has been dominated in recent years by two closely linked themes. The first is the control of disease, especially the impact of major epizootic outbreaks upon African domestic livestock production;[1] the second is the development, or to be more precise, the lack of development of commercial livestock production.[2] The connection between these two themes was the principal concern in the development of Western veterinary medicine in Africa throughout the twentieth century and especially during the colonial period up until the early 1960s. The veterinary departments established throughout colonial Africa sought to contain disease in order to bring development. This "biological warfare" was championed, as Shaun Milton has reminded us, "as part of the wider struggle of the forces of the human enlightenment over those of darkness and ignorance in the face of a merciless nature."[3] Milton's imagery reflects the "colonial mission" of the early twentieth century but also echoes the beliefs and attitudes that were most evident in the colonies of white settlement, where European immigrants struggled to establish themselves as dairy and beef producers alongside indigenous African herders. It was in the settler states of Kenya, Rhodesia, and South Africa that veterinary authorities were most aggressive in promoting a Western model of commercial development for the livestock sector. In that struggle, the connection between disease control and commercial development was absolutely crucial, as the advocates of settler production stigmatized and denigrated indigenous African producers in their efforts to promote their own interests. This was, indeed, commonly presented as a battle between the progressive, economic, and sustainable methods of production advocated by European settlers and the backward, uneconomic, and environmentally damaging practices of African herders.[4]

This chapter examines the history of the development of Kenya's livestock industry over the period from the end of World War I to 1948, precisely focusing on the tensions that emerged around the development of a settler beef industry. Taking up themes first considered for Southern Rhodesia (by Phimister[5]) and for South Africa (by Milton[6]), both examples in which state subsidy and direct regulation allowed the development of a settler-led export markets, the Kenyan story had a less happy outcome for the advocates of settler production. Lacking the political authority of their counterparts in Southern Rhodesia, or even in South Africa, Kenya's settlers struggled to win support for their plans.[7] It was only after several years of negotiation in the mid-1930s that an agreement was brokered with the Liebig company linking the development of meat processing in Kenya with a campaign to cull African-owned "scrub stock," thus neatly coupling the advance of settler production with the solving of the supposed "problem" of African livestock production. The "solution" turned into a comical farce, as market forces exposed the frailty of the European scheme. Kenya's settlers did not get their export market, nor did they succeed in bringing the African domestic livestock sector under their control.

The chapter begins by reviewing the European view of African herding in the interwar years, explaining views about African overstocking and its consequences and cures. The next section moves on to consider the promotion of the European export market and the involvement of Liebig in Kenya, cataloging the abysmal and embarrassing failure of the culling scheme put in place to feed the Liebig meat-processing plant. The chapter concludes with a brief discussion of the character of the African livestock market and its response to price incentives. The argument challenges notions of uneconomic indigenous African production, suggesting that the state's unwillingness to give Africans a fair price for their stock was the problem, not the reluctance of Africans to bring stock to the market.

The Overstocking Debate

In all the many debates about the development of Kenya's livestock industry in the interwar years, the question of overstocking emerges as a dominating theme. European opinion in the colony, both settler and official, advocated the destocking of the African areas, some favoring the provision of market incentives, others promoting outright compulsion. Debates on destocking in Kenya frequently drew upon South African examples, citing at length the Drought Commission of 1921 and the reports on the Native Economic Commission of 1930–32.[8] Kenya's own Agricultural Commission, under the chairmanship of Sir Daniel Hall in 1929, had given lengthy consideration to overstocking, stressing the urgency of the problem and the need for a formal government culling scheme.[9] Three years later, Kenya's settlers told the same tale to the Land Commission as they had to the Agricultural Commission, lamenting the economic waste of African herders who kept unproductive animals in excessive numbers. The chief veterinary officer, Major Brassey-Edwards, and his deputy, Capt. Mulligan, were among many commentators who gave evidence to the Land Commission. Both these veterinarians were sympathetic toward African herders, but they placed emphasis on the need to realize an economic return from African-owned cattle.[10] They were among the many witnesses who argued for the establishment of a meat factory capable of dealing with lower-quality animals. With a sharp eye on the costs of infrastructure development in the African areas, Brassey-Edwards wanted revenue from culling to be put back into the development of the pastoral lands—a sentiment that was not shared by the settler witnesses who gave evidence before the Land Commission.[11]

The European witnesses giving evidence before the Land Commission lamented the low quality of African livestock and poor African herding practices, but their primary concern was the threat of diseases spreading from African-owned herds to European livestock. Their aim was to see the government impose closer regulation over African herders, with compulsory disposal of diseased or "uneconomic" stock. According to European-settler opinion, the Africans held irrational attitudes toward the accumulation of cattle. Supposedly untroubled by the constraints of ecology and unmoved by the economic considerations that would determine stock management on a European model, the African herder was portrayed as driven only by cultural beliefs and social practices. Livestock ownership, and especially cattle ownership, was equated with wealth and prestige and was particularly important in relation to the acquisition of wives through the payment of a bride price. This situation distorted herding practices and led directly to overstocking, as herders accumulated cattle in order to meet social obligations. Furthermore, common-property rights in land took away any need for a constraint upon numbers. It was thus the aim of every African herd owner to accumulate as many animals as possible, regardless of the environmental consequences or the economic implications.[12]

This stereotype of the African herder had gained academic credibility in the mid-1920s with the publication of M. J. Herskovits's influential series of articles on "the cattle complex" in East Africa.[13] While Herskovits was careful to set his arguments within a cultural framework, his ideas provided a scholarly gloss to what was already a widely accepted European explanation for apparently aberrant behavior. For those European settlers who thought that the accumulation of too many cattle was symptomatic of the backwardness of African society, of its adherence to "custom" and "tradition," and of its innate conservatism, Herskovits appeared as an ally. And Herskovits's views were repeated, albeit in partial and slightly garbled form, in the writings of eastern Africa's leading veterinarians.[14] Such views informed European opinion in Kenya, were repeated before the land commissioners time after time, and by the 1930s, had become an accepted orthodoxy.

At the nub of the overstocking debate was the relationship of livestock numbers to land availability and human population, but the basic statistics for such calculations were not available for very many parts of Kenya. The evidence for overstocking, such as it was, was based in anecdote and observation. The first reasonably reliable human-population census was not conducted in the colony until 1948, and estimates of livestock holdings were seldom undertaken in any systematic manner, and then only in specific localities.[15] The truth was that all estimates of livestock numbers were nothing more than calculated guesses.

Livestock trading was also a highly contentious issue in relation to the overstocking debate. Witnesses before the Land Commission could not find agreement when it came to explaining the apparent reluctance of Africans to trade their livestock, but they were sure that that herders simply would not market their animals, no matter what the price: irrationality and cultural values prevented their behaving in an economically sensible manner.[16] A small minority of commentators, including several senior officers in the political administration, among them the Rift Valley Provincial Commissioner H. E. Welby, suggested that this problem was accentuated by veterinary policies. Government-imposed quarantines, put in place to protect European-owned herds from disease, "locked up" the African reserves like "little tin-boxes," Welby argued, preventing the legal export of animals.[17] Because of these policies, for much of the time, most African herders could not legally sell livestock for export from their reserve even if they wanted to. Overstocking was, at least in part, the product of the policy of quarantines, Welby contested. He wanted quarantines lifted and stock routes established so that a regular outlet could be provided for African stock. Only then, he argued, could an effective program of destocking be implemented, whether compulsory or voluntary.

European settlers vehemently opposed this view, claiming that it would result in the destruction of the settler dairy and beef industries. The efficacy of the "tin-box" policy had, in fact, been debated periodically since the early 1920s, with those who wished to see the development of the African reserves arguing for a change and those wishing to protect settler dairy and beef producers seeking to maintain the status quo.[18] Even among veterinarians, there were some who doubted the value of quarantines without accompanying disease-eradication programs. These arguments were also rehearsed before the land commissioners between August 1932 and May 1933.[19] Among the commissioners themselves, there was dispute over the matter, one (Wilson, a Kenya farmer and stockowner who adopted the settler "firm line") clearly favoring the maintenance of quarantine controls and compulsory culling even if uneconomic, with another (Hemsted, a former member of the political administration) supporting the need to develop markets for African stock. The argument was ultimately won by Wilson, who took responsibility for drafting the parts of the final report dealing with overstocking. The land commissioners, perhaps predictably enough, came down on the side of protecting settler economic interests: the quarantines were needed, it was concluded, but steps should be taken to establish a factory for processing low-quality African cattle that might be acquired through controlled culling measures.[20] Behind this conclusion, which repeated a recommendation of the Agricultural Commission of 1929, lay the basic assumption that Africans would not willingly market livestock and that even if they could be persuaded to do so, it would only be the poorest stock that would be offered for sale.

By the mid-1930s, colonial officials in London shared in these views of irrational and uneconomic African herding. The advisor to the Colonial Office on veterinary and agricultural matters, Frank Stockdale, was firmly of the opinion that no progress could be made in Kenya's rural development until the livestock kept by Africans were dramatically reduced by culling. Whether a cull of African-owned livestock could be imposed had been vigorously debated within Kenya over a good many years. Even before World War I, there had been proposals to introduce a cattle tax in the Maasai districts, and during the war, the government had first investigated the possibility of establishing a meat-canning factory to dispose of the livestock that might be brought to market if a tax was to be introduced.[21] This proposal resurfaced in different forms on several occasions in the 1920s, most notably in 1927, when local officials put forward a scheme for the compulsory culling of cattle in the Kamba areas of Machakos, to the east of Nairobi. Along with the Baringo district in the northern Rift Valley, Machakos was considered to be among the most heavily overstocked of Kenya's African landscapes. Although the plan for compulsory culling was energetically supported by Governor Grigg, the idea was squashed by the Colonial Office, "which feared Kamba unrest and favored a more gradual approach."[22] Despite these anxieties, the idea was floated again in the Agricultural Commission in 1929, with the firm suggestion that a canning or processing plant for African scrub stock should be constructed.[23] On this occasion, the Veterinary Department showed enthusiasm, and inquiries were made as to the costs and likely commercial implications of such a development. But the commission's report proved to be untimely, as the Great Depression descended and the colonial economy entered a period of retrenchment.

The Land Commission then revived the idea of a meat-processing and canning plant in 1933, but its pessimism about the low quality of the African-owned stock likely to be offered for sale led the commissioners to advocate the establishment of a fertilizer factory as a more appropriate means of disposing of the surplus cattle. This proposal had the advantage of being affordable without the involvement of a commercial company, and so a proposal to build such a factory was consequently put to the Colonial Development Fund. A grant of 23,590 pounds was received in 1936. However, in making the grant, officials in London insisted that the fertilizer scheme should be linked to the idea of progressively educating the African herder as to the need to rid himself of poor-quality stock, and not to any predetermined program of culling.

Liebig and Compulsory Culling

This was not the solution that Kenya's white settlers had wanted, for while a fertilizer factory offered a solution to the problem of what to do with low-quality scrub stock, it did not provide the canning and freezing facilities that they required in order to develop an export market for their beef. They found an ally in Kenya's new director of Veterinary Services, R. Daubney, who was appointed after the publication of the Land Commission report. Daubney revived the notion of obtaining commercial investment for the beef industry and opened negotiations with the Liebig company. In 1934,

Liebig had established a meat processing, freezing, and canning operation in Southern Rhodesia. Supported by state regulation of the livestock market, this had proved a successful model, and in June 1936, Daubney visited Rhodesia to see the Liebig factory for himself. He returned brimming with enthusiasm for the potential of canning and freezing in Kenya and immediately opened negotiations with Liebig. Before the month was out, a senior representative of the company was in Kenya to assess the market opportunities, and by August 1936, draft terms and conditions for a proposed factory on the railway at Athi River, to the southeast of Nairobi, were under discussion.[24]

Daubney did not inform London of his plans until he had a clear view of the terms that Liebig would require to make a Kenyan factory viable. When the proposal was finally submitted to the Colonial Office, emphasis was placed on the veterinary aspects of the scheme and especially on the prospects for developing an export market of high-quality meat—produced from settler-owned herds—to the British market. The Colonial Office gave its support to the proposal in principle, while pointing out that the question of importing Kenyan meat to the United Kingdom would be difficult to resolve because of the issues of disease control and import quotas governed by the Board of Trade. Though these obstacles were considerable, the Kenyan administration pushed ahead with the scheme. Liebig was given land for its factory and a ten-year lease on ten thousand acres for a holding ground, with an option on a further ten thousand acres adjacent to the factory site. The colonial government also agreed to "afford all reasonable facilities" for the movement of stock to the Athi River site, thereby committing itself to the creation of stock routes, and gave assurances that "everything possible" would be done to secure the supply of cattle to the factory. These were promises the Kenya government would find it difficult to keep.

Liebig estimated that the Athi River factory would process up to thirty thousand head of cattle per annum, purchasing the stock at between Shs 2/50 to Shs 4/- per one-hundred-pounds live weight. By Liebig's calculations, this gave the herder a "fair price" of between Shs 18/- and Shs 30/- for an animal of 750 pounds.[25] This price was "fair" only in the sense that Kenya's livestock trade was then in an economic slump, with prevailing prices much lower than their longer-term average.[26] Daubney was confident that Kenya could supply stock in sufficient quantities at these prices to make the factory viable, although it was evident that a marketing infrastructure would have to be hastily assembled. With Daubney pushing the scheme forward, the grant initially received for the fertilizer factory was canceled, and a sum of 11,400 pounds received in its place for the development of stock routes.[27] Leaving the Kenya government to worry about supply, Liebig set about building its factory, which was ready to open early in 1938.

Frank Stockdale would later remark that Daubney had been "a supreme optimist" in imagining all this would be possible.[28] It was Daubney's hope that the factory would be supplied through culling campaigns mounted in the African reserves, yet he never squared up to the political implications of this. The Rift Valley, including Baringo, the Northern Frontier Province, the Maasai districts, and Machakos had all been mentioned in discussions with Liebig as likely areas of supply, and it was on the market prices prevailing in these areas that Liebig had based its costings in 1936. However, by the time the factory was opened in 1938, the market in these areas had largely recovered, and Liebig found it difficult to buy cattle at even double the earlier estimated prices.[29] Unable to make a profit if it paid the higher prices, Liebig immediately experienced a shortfall of supply. Only a few months after the opening of the Athi River plant, the company threatened to close down the operation unless the government assisted in securing the supply of cattle through compulsory culling.[30]

Although there is no evidence that Daubney favored compulsion from the outset, he had always expected the provincial commissioners to support the factory by making every effort to secure cattle from the pastoral areas. Despite the fact that most of these senior officials remained opposed to compulsion, in December 1937 Daubney succeeded in pushing through a decision to mount an experimental compulsory destocking campaign in Machakos, to the east of Nairobi.[31] The Kamba area of Machakos was among the most notoriously overstocked of Kenya's African reserves.[32] After several years of work on antierosion measures in Machakos, a "Reconditioning Committee" had been established in 1935, comprising local chiefs. Propaganda in favor of destocking was issued through the Reconditioning Committee. Of more significance, however, the administrative officers in Machakos were generally more persuaded of the need for compulsion than were their counterparts in other districts. District Commissioner A. N. Bailward was instructed to draw up a scheme for destocking that would allow the animals to be sold in lots at public auction, by which means prices would be held down to a level that Liebig could afford. To achieve this, Bailward set culling quotas for each locality, based on estimated carrying capacity, and left it to the local chiefs and headmen to decide which cattle would be sacrificed to the cull. By wooing Liebig to Kenya, Daubney had, in effect, forced the issue of compulsory culling to a head after more than a decade of official prevarication.

As the destocking campaign got underway, the settler press was brimming with self-satisfied delight: here was the culmination of more than a decade of settler pressure upon the government, and once again the government had been forced to concede that the settler, in fact, knew best.[33] The initial phase of the cull appeared to proceed well, and stock began to arrive at the Liebig factory. But before the end of July, the whole program had been derailed by African protest, with two thousand Kamba marching to Nairobi to confront the governor and widespread refusal to cooperate with the chiefs and headmen seeking to organize the cull in the reserve. While settlers called for strong measures to force compliance, the government dallied and eventually backed down: the cull was first postponed, then canceled, the placatory response urged by officials in London who suggested that Nairobi adopt a policy of "encouragement," rather than compulsion.

The successful Kamba protest placed the entire Liebig enterprise in jeopardy. Without compulsion, African herders would not offer sufficient cattle to Liebig's buyers at the low prices they were willing to pay. After stumbling through several more difficult months, the parties involved found that, by April 1939, supplies had all but dried up, and Liebig took the decision to close the Athi River factory.[34] Under a barrage of settler criticism, led by the Stock Owners' Association, Daubney sought concessions that would allow the factory to reopen. With drought then affecting the highlands, settlers offered to supply the factory with a limited number of stock from their own herds. In addition, Daubney worked hard to open up further stock routes from the north and the Rift Valley. But none of this solved Liebig's problem.

Having been failed by the Kenya government, Liebig looked elsewhere for its supply. Cheaper cattle could be bought in Tanganyika and even from Uganda, and Liebig now requested permission to import stock from these neighboring territories. The Kenyan administration reluctantly capitulated, agreeing to unlimited imports for six months and undertaking to underwrite any losses made through shortfall of supplies in that period.[35] By November, Daubney's plan to bring cattle from the northern Rift Valley had been thwarted by the diagnosis of bovine pleuropneumonia among the Samburu cattle: the factory was fast becoming totally dependent upon imported stock. A dejected Daubney attended the meeting of provincial commissioners on 13 November 1939, in hope of persuading them "that by the exercise of a little pressure" more Kenya cattle could be supplied to the factory.[36] With their fingers so recently burned in Machakos, no provincial commissioner was prepared to offer any assistance to Daubney.

By the end of 1939, Liebig's management reported that "the position has now developed into the ludicrous one that we have a factory in Kenya and that every animal we shall be slaughtering between now and the New Year will be coming from a neighboring territory."[37] Before the year was out, a steady supply of cattle was entering Kenya from Tanganyika, utilizing Kenya's newly provisioned stock routes to make its way to the Athi River factory. During 1939, the market price of cattle in Tanganyika had fallen steadily, to a level at which Liebig's buyers could compete effectively with Somali and other African traders at the regular markets held in several provinces. Here, without the restrictions imposed by the disease controls that protected settler dairy and beef producers in neighboring Kenya and without any compulsion, there was a buoyant African livestock trade that saw more than one hundred thousand head of cattle sold through the official markets each year.[38]

In effect, the Kenya government was now subsidizing the destocking of Tanganyika's pastoral areas. From its reopening in July 1939 until the end of the year, the Athi River factory handled 11,792 head of stock, 3,846 being imported from Tanganyika and the majority of the remainder coming from settler farms in the White Highlands. Over the next three years, the trend was accentuated: in 1940, the factory handled 58,044 cattle, 48,038 of these from Tanganyika; in 1941, the total increased to 86,414, of which 67,769 head came from Tanganyika; and in 1942, 63,465 of the total of 77,537 head of stock processed by the Liebig company were imported from Tanganyika.[39] As the absurdity of the situation unfolded, Stockdale could not hide his disappointment at the failure but supplied minutes to

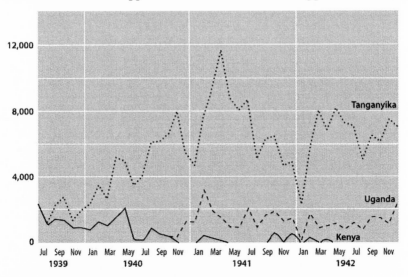

Figure 14.1. Livestock supply to Liebig (Athi River) July 1939–December 1942

his Colonial Office colleagues indicating that he felt "the Kenya govern-ment did not think out the question of supply carefully enough."[40]

The Livestock Market

In the midst of the humiliating setbacks with Liebig, the Kenya government set up an "Overstocking Committee" to make recommendations for a new set of policies. The committee of seven included four settler members, two of whom, Colvile and Pardoe, were appointed to represent the Stock Own-ers' Association. At the opening meeting, on 29 May 1939, Colvile made it clear that his association believed compulsory culling was the only way forward.[41] At the second meeting, on 15 June, the two men forced the issue again, insisting that the government compel African herders to comply with culling orders. When the government refused to give any such assurance, the settler members resigned from the committee.[42] With their departure, the debate over compulsion was effectively closed. When the committee produced its interim report in April 1941, the emphasis was placed upon the creation of a livestock market of the kind operating in Tanganyika, with formal and regular stock auctions in the African reserves. Stock routes would be opened by the Veterinary Department from these markets to give an outlet for the cattle; education, propaganda, and persuasion were to be the weapons to combat overstocking; and while the possibility of culling was not excluded, it was only to be introduced with the cooperation of the local African authorities.[43] This was to be the basis of Kenya's postwar development policy for the livestock sector.[44]

The investigations of the Overstocking Committee identified the price offered to the African herder as the crucial element in determining the level of market engagement. This was brought into sharp relief by the re-markably successful activities of the Meat Control Board, set up in Kenya after the outbreak of war to secure beef supplies for the military. In the six months between September 1941 and March 1942, the meat control-ler purchased eighty-nine thousand head of cattle from Kenya's African reserves. Over the same period, Liebig was able to secure only 1,842 head of cattle from the same areas. The crucial difference was price: Liebig was offering only Shs 5/- per 100 pounds weight for first-grade animals, Shs 3/50 for second grade, and Shs 3/- for third grade; the Meat Control Board paid Shs 10/- for first grade, Shs 7/- for second grade, and Shs 4/- for third grade (though it rarely bought animals below the second grade).[45] Liebig complained that the activities of the Meat Control Board, supported as it was by the weight of the state, amounted to requisitioning. There was more than a grain of truth in this, and before the end of the war, African opposition

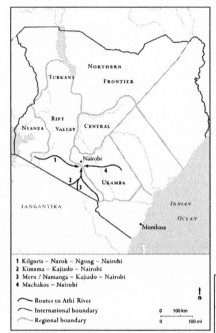

1 Kilgoris – Narok – Ngong – Nairobi
2 Kimama – Kajiado – Nairobi
3 Meru / Namanga – Kajiado – Nairobi
4 Machakos – Nairobi

↑ n

⌒ Routes to Athi River
⌒ International boundary
⌒ Regional boundary

0   100 km
0   100 mi

Figure 14.2. The development of stock routes, 1940 and 1948

1 Nyeri – Thika – Nairobi
2 Kipsigis to railhead
3 Kitui to Nairobi
4 Machakos to Nairobi
5 NFP: Isiolo – Nanyuki
6 Mukugodo to Isiolo
7 Samburu to T. Falls
8 Elgeyo – Marakwet to Eldoret
9 Elgeyo – Marakwet to Timboroa
10 Nandi to Eldoret
11 Baringo: Mogotio to McCall's Siding
12 Turkana & W. Sok: Kapenguria – Kitale
13 Nyanza: road to Eldoret; rail to Nairobi
14 Maasai: Kajiado & Tanganyika stock
15 Maasai: Kilgoris – Narok – Ngong – Nairobi

↑ n

0   100 km
0   100 mi

⌒ Routes to Athi River
⌒ International boundary
⌒ Regional boundary

to the Meat Control sales was becoming apparent in some districts. The Stock Owners' Association now worried that the sale of higher-quality animals to the board was diminishing the breeding stock of the herds without dealing with the more immediate problem of poorer scrub stock. But the clear message was that African herders, far from being "uneconomic" or "irrational" in their attitudes to livestock, were highly sensitive to price shifts in the market: they sold stock when they judged the price to be favorable and held on to it when prices were poor.

Studies of livestock marketing from many parts of East Africa now show that African herders tend to market cattle irregularly, but that at times this trade can be vigorous.[46] Throughout pastoral areas of the Rift Valley, livestock trading had been an important factor in the development of herds from the early 1900s up to the 1920s. Trading was always especially vigorous in the wake of drought periods, as herders traded back up from sheep and goats into cattle. At these times, large numbers of small stock came to the market, and the prices of cattle would rise very sharply. This market was facilitated by itinerant traders. Even before 1910, itinerant Somali livestock traders were a common sight throughout the Rift Valley, as cattle bought cheaply in the north were moved south to take advantage of better prices. European settlement itself gave a stimulus to this trade, with the indigenous traders supplying settlers with stock. During World War I, the government then encouraged an increase in itinerant livestock trading as a means of securing a meat supply for the military; throughout the 1914–18 war, Somali traders supplied government buyers at Nakuru with animals, for example, many of these bought in the Baringo district.[47] By the 1920s the itinerant Somali traders had been joined by Nubians and a growing number of other traders, including Kikuyu.[48]

Government quarantine regulations increasingly influenced the pattern of stock trading from 1918. The need to control the spread of rinderpest, bovine pleuropneumonia, and other cattle diseases placed an almost permanent prohibition on the legal movement of cattle from the African pastoral reserves through the settler farmlands. However, the quarantines imposed did not prevent the movement of potentially diseased stock into African areas close to the settler farms from more remote northern locations.[49] Cattle could be purchased in the northern Rift Valley for around one-third the equivalent cost in the area around Nakuru to the south. During the early 1920s, a trader might purchase a heifer in Turkana for the equivalent of about Shs 15/-.[50] This animal would then be trekked south, into Baringo, where it would be exchanged for goats. Before 1930, a trader might expect to receive as many as thirty decent-sized goats in exchange for a single heifer. After 1930, this figure gradually decreased, falling to around twenty by 1939. At the markets in the settled areas of Rongai, Njoro, Nakuru, and Subukia, Shs 3/- to 5/- per goat could be obtained from Kikuyu buyers purchasing meat for the butcheries of Nakuru and Nairobi. The heifer purchased for Shs 15/- in Turkana had, therefore, realized perhaps over Shs 100/-.[51]

But while it is clear that there was a lively trade in livestock and livestock products throughout the interwar years, it is very difficult to specify what numbers of stock were annually exchanged in any particular district. Government figures for sheep and goats officially leaving Baringo under export permit, for example, vary dramatically between peaks of over eighty thousand head during World War I to lows of just over ten thousand per annum during the drought years between 1928 and 1932. These figures include animals "in-transit" from areas further north, amounting to perhaps 60 or 70 percent of the totals in some years.[52] But it seems certain that the official government figures represent only a small portion of the actual export trade and that a considerable amount of internal trade never came to official notice. The problem, then, was never that Africans would not bring stock to market, as Kenya's settlers claimed, but rather that the livestock market could not be managed in a manner that facilitated the development of the European sector without the support of state intervention and regulation.

THE STORY of Kenya's cattle trade between 1918 and 1940 highlights the dynamism of African agency in responding to market opportunities and the uninvited challenges that colonialism presented. When there was a market and a fair price, African pastoralists were willing to trade, thereby confounding European assumptions that Africans attributed only a cultural value to their animals and would not participate in the livestock economy unless coerced. The European obsession with a supposed African "cattle complex" was indicative of the divisions between the colonizers and colonized in Kenya, marked by competition between African producers and their European settler rivals for land, labor, and state support, and it underlined the extent of British misperception of the character of African husbandry practices.

This episode also demonstrates the fragility of British state policies on the ground. Africans were often successful in defying British attempts to reform animal husbandry and rejected claims that the land was becoming degraded through overstocking. Kenya's African pastoralists refused to reduce the size of their herds voluntarily and to supply the Liebig company with meat just at the behest of the colonial government, unless it was profitable to themselves, which it clearly was not. Kenyan herders, instead, found ready buyers among Somali traders, and the British proved unable to manipulate the African economy to suit their interests. Fears of generating rural unrest by enforcing unpopular initiatives such as compulsory destocking further reduced the ability of the British to manage the rural economy, leading to loud complaints from European settlers that the colonial government lacked the will to implement a coherent policy. There were, after all, limits to the extent to which the British government was prepared to underwrite a settler-dominated agricultural economy in Kenya.

The peculiar history of Liebig in British colonial Africa also reminds us of the importance of examining the "local" when assessing the economics of empire. Liebig was able to secure a successful meat market in Southern Rhodesia and found an unintended source of supply for its Kenyan factory from Tanganyika's lively stock trade. But within Kenya, British attempts to encourage the development of an African commercial livestock sector by dictating both the terms of production and trade were less than successful. Kenyan herders were able to take advantage of gaps in the veterinary cordons and the profusion of provincial markets to ensure that they profited at the cost of the settler state.

Notes

1. For example, Charles van Onselen, "Reactions to Rinderpest in Southern Africa, 1896–97," *Journal of African History* 13, no. 3 (1972): 473–88; R. T. Wilson, "The Incidence and Control of Livestock Diseases in Anglo-Egyptian Sudan, 1916–56," *International Journal of African Historical Studies* 12, no. 1 (1979): 62–82; J. L. Giblin, "East Coast Fever in Socio-historical Context: A Case Study from Tanzania," *International Journal of African Historical Studies* 23, no. 3 (1990): 401–21; J. L. Giblin, "Trypanosomiasis Control in African History: An Evaded Issue?" *Journal of African History* 31, no. 1 (1990): 59–80; Richard Waller, "Tsetse Fly in Western Narok, Kenya," *Journal of African History* 31, no. 1 (1990): 81–101; J. M. Mackenzie, "Experts and Amateurs: Tsetse, Nagana and Sleeping Sickness in East and Central Africa," in *Imperialism and the Natural World*, ed. J. M. Mackenzie (Manchester: Manchester University Press, 1990), 187–212; William Beinart, "Vets, Viruses and Environmentalism at the Cape," in *Ecology and Empire: Environmental History of Settler Societies*, ed. T. Griffiths and L. Robin (Keele: Edinburgh University Press, 1997), 87–101; and Ian Scoones and William Wolmer, "Land, Landscapes and Disease: The Case of Foot and Mouth in Southern Zimbabwe," *South African Historical Journal* 58 (2007): 42–64.

2. For a discussion, see David Anderson, "Cow Power: Livestock and Pastoralists in Africa," *African Affairs* 92, no. 366 (1993): 121–33. For relevant examples, see Peter L. Raikes, *Livestock Policy and Development in East Africa* (Uppsala: NAI, 1981); Carol Kerven, *Customary Commerce* (London: ODI, 1992); and Fred Zaal, *Pastoralism in a Global Age: Livestock Marketing and Pastoral Commercial Activities in Kenya and Burkina Faso* (Amsterdam: Free University, 1998).

3. Shaun Milton, "Western Veterinary Medicine in Colonial Africa: A Survey, 1902–1963," *Argos* 18 (1998): 313.

4. Ian R. G. Spencer, "Pastoralism and Colonial Policy in Kenya 1895–1929," in *Imperialism, Colonialism and Hunger in East and Central Africa*, ed. R. I. Rotberg (Lexington, KY: Lexington Books, 1984), 113–40, discusses this point. For an example of the "procolonial" genre of veterinary history, see K. D. S. MacOwan, "The Development of a Livestock Industry in Kenya," *Veterinary History* 8, nos. 1–2 (1994): 29–37.

5. Ian Phimister, "Meat and Monopolies: Beef Cattle in Southern Rhodesia, 1890–1938," *Journal of African History* 19, no. 3 (1978): 391–414.

6. Shaun Milton, "The Transvaal Beef Frontier: Environment, Markets and the Ideology of Development," in Griffiths and Robin, *Ecology and Empire*, 199–214; and idem, "The Apocalypse Cow: Russell Thornton and State Policy toward the African Cattle Industry in the Union of South Africa, c. 1924–39" (paper presented to the African History Seminar, SOAS, London, November 1994).

7. Regarding South Africa, see Shaun Milton, "To Make the Crooked Straight: Settler Colonialism, Imperial Decline and the South African Beef Industry" (PhD diss., University of London, 1996).

8. For examples, all from *Kenya Land Commission: Evidence and Memoranda (KLC:EM)*, vol. 3, "Memorandum from the Chief Native Commissioner," 3087–91; "Report of the Board of Agriculture on Overstocking," 3091–97; S. H. La Fontaine, "Memo: Overstocking in Native Reserves," 3101–3; evidence of H. H. Brassey-Edwards (chief veterinary officer), 3103–13; and memo by Brassey-Edwards and E. J. Mulligan, "Overstocking of Native Reserves and Disposal of Native Stock," 3114–19.

9. *Report of the Agricultural (Hall) Commission* (Nairobi: Government Printer, 1929).

10. For a fuller discussion of the Land Commission, see David Anderson, *Eroding the Commons: Politics and Ecology in Baringo, Kenya* (Oxford: James Currey, 2002).

11. *KLC:EM*, vol. 3, evidence of Mulligan, 3119–26.

12. For the southern African variations on this narrative, see Ian Scoones, "Politics, Polemics and Pasture in Southern Africa," and William Beinart,

"Environmental Destruction in Southern Africa," both in *The Lie of the Land: Challenging Received Wisdom on the African Landscape,* ed. Melissa Leach and Robin Mearns (Oxford: James Currey, 1996), 34–72.

13. M. J. Herskovits, "The Cattle Complex in East Africa," *American Anthropologist* 28 (1926): 230–72, 361–88, 494–528, 633–64.

14. H. E. Hornby, "Overstocking in Tanganyika Territory," *East African Agricultural Journal* 1, no. 5 (1935/36): 355–66

15. For typical examples of livestock censuses, these for the northern Rift Valley, see Rimington (DC/Baringo) to PC/RVP, 2 April 1931, PC/RVP.6A/23/l, Kenya National Archive (KNA); McClure (stock inspector) to PC/RVP, 17 September 1936, "An Analysis of the 1936 Stock Census, in Relation to the Land and the People," PC/RVP.6A/23/lo, KNA.

16. For example, *KLC:EM,* vol. 2, comments of Stanning, Dwen, and Lean, 1820–28.

17. *KLC:EM,* vol. 3, evidence of Welby, 3142–47.

18. Spencer, "Pastoralism and Colonial Policy," 113–40; Raikes, *Livestock Policy and Development,* 19–22.

19. For the most relevant comments by veterinary officers, see *KLC:EM,* vol. 3, evidence of Brassey-Edwards, 3103–14; evidence of Brassey-Edwards and Mulligan, 3114–19; evidence of Mulligan, 3119–26; and evidence of McCall, 3167–74.

20. *KLC: Report.*

21. See the report by Ainsworth of 26 April 1915, CO 533/154, British National Archives (BNA), and the memorandum by Hemsted of 30 June 1918, enclosed with Bowring to Long, 13 March 1918, CO 533/194, BNA.

22. J. F. Munro, *Colonial Rule and the Kamba: Social Change in the Kenya Highlands, 1889–1939* (Oxford: Clarendon Press, 1975), 220, citing correspondence in CO 533/372/10510, BNA.

23. *Report of the Agricultural Commission (Hall),* passim.

24. De Wade to Ormsby-Gore, 12 September 1936, and subsequent correspondence, CO 852/16/8, BNA.

25. De Wade to Ormsby-Gore, 12 September 1936, CO 852/16/8, BNA.

26. See Anderson, *Eroding the Commons,* 202, for statistics on prices.

27. De Wade (ag. gov.) to Ormsby-Gore, 20 March 1937, and Ormsby-Gore to Morrison (Board of Trade), 24 July 1937, both in CO 852/105/9, BNA.

28. Minute by Stockdale, 21 December 1939, CO 852/219/12, BNA.

29. Liebig did manage to buy stock in Baringo, although the quality was very poor: Dir. of Vet. to PC/RVP, 3 August 1939, and reply 1 September 1939, PC/RVP.6A/23/13, KNA.

30. Secretariat Circular, no. 23, 20 December 1937, CO 533/496/38184/3, BNA, cited in Munro, *Kamba,* 221.

31. Munro, *Kamba,* cites a note of a discussion held at Government House, 21 December 1937, in VET 2/10/7/1 II, vol. 2, deposit no. 1/23, KNA. I have been unable to locate this file.

32. The following account summarizes Munro, *Kamba,* 220–22; Robert L. Tignor, "Kamba Political Protest: The Destocking Controversy of 1938," *International Journal of African Historical Studies* 4, no. 2 (1971): 237–51; and Bismarck Myrick, "Colonial Initiatives and Kamba Reaction in Machakos District: The Destocking Issue, 1930–38," in *Three Aspects of Crisis in Colonial Kenya,* ed. L. Spencer, Foreign and Comparative Studies Series, East Africa no. 21 (Syracuse, NY: Syracuse University, 1975), 1–26.

33. For a sample from the *East African Standard,* see "Destocking Sales Continue," 4 June 1938, 28; "Auctioning Kamba Cattle," 18 June 1938, 7; "Kenya and Export of Meat: New Chilling Factory May Pave Way," 21 June 1938, 5; and "Impression of a Destocking: Distribution Rather Than Destruction," 1 July 1938, 3.

34. Minutes by Melville, 30 May 1939, and Williams, 30 December 1939, both BNA, CO 852/219/12, BNA.

35. Brooke-Popham to sec. of state, 29 August 1939, CO 852/219/12, BNA. See also *East African Standard,* "Subsidy for Liebig's Suggested," 28 April 1939; "Plan to Help Kenyan Stock Industry," 30 June 1939; and "Liebig's Reopening," 7 July 1939.

36. Daubney to Stockdale, 12 November 1939, CO 852/219/12, BNA.

37. Liebig's manager, Nairobi, to Liebig's Oxted, England, 22 November 1939, copied in CO 852/219/12, BNA.

38. "Cattle Walk 450 miles to Kenya," *East African Standard,* 19 December 1939.

39. Calculated from the Athi River slaughter returns.

40. Minute by Stockdale, 4 June 1940, CO 852/288/14, BNA.

41. "Note of a Discussion of the First Meeting of the Overstocking Committee," 29 May 1939, CS 1/5/1, KNA.

42. "Overstocking Committee Circular," to all members of the Executive Council, 15 June 1939; Colvile to ag. chief secretary, 23 June 1939; Pardoe to ag. chief secretary, 24 June 1939, all in ARC(MAWR)-3 Vet-1/8, KNA.

43. *Interim Report of a Committee Appointed to Advise as to the Steps to Be Taken to Deal with the Problem of Overstocking in Order to Preserve the Future Welfare of the Native Pastoral Areas* (Nairobi: Government Printers, 1941). For the interim report and related papers, see ARC(MAWR)-3 Vet -1/13ii, KNA.

44. Raikes, *Livestock Policy and Development,* 193–96

45. The prices were reported to the Executive Committee of the Stock Owner's Association, 30 October 1940, Min. of Agri. 2/32, KNA. Liebig raised its grade prices in April 1942, to Shs 7/60, Shs 6/-, and Shs 5/-, respectively: see "Summary of Discussions re. Livestock Purchases," 9 April 1942, ARC(MAWR)-3 Vet-3/28, KNA.

46. Raikes, *Livestock Policy and Development*, esp. 191–204, is the most comprehensive study. For historical discussion of the issue in eastern Africa, see E. Harrison, *The Native Cattle Problem*, Agricultural Department Pamphlet (Nairobi: n.p., 1929); H. R. Bisschop, *Improvement of Livestock*, Department of Agriculture, pamphlet no. 424 (Nairobi: n.p., 1949); and Michael D. Quam, "Cattle-Marketing and Pastoral Conservatism: Karamoja District, Uganda, 1948–1970," *African Studies Review* 21, no. 1 (1978): 49–71. For Rhodesian comparisons, see Murray C. Steele, "The Economic Function of African-Owned Cattle in Colonial Zimbabwe," *Zambezia* 9, no. 1 (1981): 29–48; and R. M. G. Mtetwa, "Myth or Reality? The Cattle Complex in South East Africa, with Special Reference to Rhodesia," *Zambezia* 6, no. 1 (1978): 23–35.

47. Baringo District Annual Reports (AR), 1914–15 and 1918–19, BAR/I, KNA; Peter Dalleo, "Trade and Pastoralism: Economic Factors and the History of the Somali of North-East Kenya, 1892–1948" (PhD diss., Syracuse University, 1975).

48. Ian R. G. Spencer, "The Development of Production and Trade in the Reserve Areas of Kenya, 1895–1929" (PhD diss., Simon Fraser University 1975).

49. Occasional quarantine rules briefly halted the export of goats, such as that of 1936 when Baringo goats were suspected of transmitting rinderpest to settler-owned stock at Njoro: Murphy (DC/Baringo) to DC/Nakuru, 17 January 1936, PC/RVP.6A/23/9, KNA.

50. Interviews with Haji adam Halijab, Ibrahim Fatamullah, and Ali bin Salim, retired livestock traders from the northern Rift Valley.

51. Oral histories collected in 1980 and 1981 gave evidence on these prices, and this evidence was then correlated with livestock prices recorded by colonial officials; see Baringo District AR, 1914–15, BAR/l, KNA; and DC/Eldoret to PC/RVP, 8 May 1937, PC/NKU/2/34/21, KNA.

52. For estimated figures, see Baringo District ARs for 1913–l4, BAR/I, KNA; for 1935, PC/RVP.2/7/3, KNA; and for 1938, PC/RVP.2/7/3, KNA.

Conclusion

Karen Brown

CONTRIBUTORS TO this volume have described a number of important case studies from Europe, North America, Africa, Asia, and Australasia. Collectively, they have explored the gradual professionalization of veterinary services as a result of developments in science and technology and the growing powers of the state, as well as the emergence of veterinary departments either in response to economic opportunities and/or the impact of devastating epizootics such as rinderpest. In addition, some of the authors have looked at the initiatives of farmers and pastoralists whose understandings of the disease environment were and continue to be based on individual observation, backed by generations of practical experience in the field. At times, local knowledge was at odds with the aims and directives of the official veterinary establishment. For many livestock owners, veterinary incursions were deemed of little use unless they resulted in the ostensible saving of animal lives, produced a notable increase in profits, or were compatible with existing agricultural and labor practices. With the exception of the chapters by Dominik Hünniger and Peter Koolmees, the historical time frame has been heavily centered on the nineteenth and twentieth centuries, when there was a shift in Western approaches to the etiology of diseases, brought about by the growing ascendancy of germ theories. These changes occurred contemporaneously with the expansion of centralized political control in Europe, North America, and some European colonies, facilitating the emergence of veterinary departments as adjuncts of modernizing states. The chapters also showed that tensions abounded between promoters of Western, technical biomedical science and guardians of folk knowledge, between governments and populace, between colonial rulers and their subjects.

Together these chapters make a significant contribution to the existing historiography on veterinary science and livestock economies, which as the introduction revealed, is rather slim. They point the way to a range of potential topics for further study and provide a baseline for comparative research on a number of diseases and themes. Drawing upon this collection and some of the recent literature in the history of human medicine, I will consider some of the many possible ways forward.

A cursory glance at the contents list alone invokes three key observations: the absence of Latin America; the prominence of rinderpest as a catalyst for veterinary interventions and reforms; and the dominance of the nineteenth and twentieth centuries. But as Hünniger and Koolmees have shown, it is possible to find documents relating to earlier periods, at least for western Europe. Further research in this field could unearth some interesting revelations about the nature and impact of disease on livestock economies, the part played by cross-border trade and warfare in the dissemination of epizootics, popular understandings and responses to animal infections, and limits to the authority of medieval and early modern states. Latin America, with its important cattle economy, has enormous potential for research. This is especially so as, along with the Caribbean, it has some unique epidemiological features, such as the transmission of paralytic rabies by vampire bats, which Rita Pemberton referred to in her contribution on Trinidad and Tobago. More work on rinderpest would be equally rewarding because of its transcontinental spread through trade, warfare, and colonialism. The rinderpest panzootic of the late nineteenth century, for example, could be explored in a global context that examines how and why diseases cross continents and how people on the spot responded to livestock crises in different parts of the world. Myron Echenberg's recent publication on the bubonic plague pandemic, which was almost concurrent with this rinderpest panzootic, mirrors some of the ideas and challenges that surrounded the spread of rinderpest and provides an interesting example of a historiographic approach that could be adapted for a monograph on a livestock disease.[1]

Moving from the contents page to the chapters themselves, a notable omission is the question of gender. This book is about men: male livestock owners and traders, male responses to livestock diseases, male scientists, not to mention male-dominated governments and veterinary departments. Far more historical research needs to be carried out into the role of women in livestock economies and how this varied from place to place and altered over time. In many African societies, for example, rural women had a very specialized knowledge of medicinal plants and might have contributed to the development of local pharmacopoeia for the treatment of animals. Colonialism also had a marked impact on the position of women in African societies. Customary taboos that existed in many communities, forbidding women from handling cattle, began to disappear in the wake of increased labor burdens due to veterinary regulations such as compulsory dipping to control tick-borne diseases, as well as male migrancy and other socioeconomic changes. Much more could be written on these issues for Africa, and doubtless an exploration of archival documents, complemented by oral testimonies for the more recent period, would reveal similarly interesting rural transformations in other parts of the world and for earlier time periods.

Along with an absence of women from the literature is the paucity of studies on animal diseases other than those affecting cattle, sheep, and horses. Pemberton and Koolmees make brief references to pig and poultry diseases for Trinidad and the Netherlands, respectively; but there is room for far more studies, especially given the close links between poultry rearing and human health. This is particularly pertinent given the situation in Southeast Asia where many people live in close proximity to their birds and some have died of the H5N1 avian influenza virus, fuelling apocalyptic fears in the media (and among some scientists) that this microbe might evolve into a far more efficient human-killing machine.

Studies too could explore the interrelationship between animal and human health more broadly. Rabies, for example, is a fascinating subject for research given the close links between humans and their pets. Rabies can also posit a major threat to local and national economies due to wildlife transmission to livestock by vampire bats in Latin America and the Caribbean, for instance, and by yellow mongooses and jackals in southern Africa. So far, accounts of rabies have been limited to Europe, but, once again, there is room for exciting comparative analysis.[2] Investigations into other infections that pass between wildlife and domestic animals, such as tick-borne diseases, malignant catarrhal fever, and tuberculosis, provide opportunities to look at shifting interrelationships between human societies and the wild and the opportunistic adaptability of microbes to new hosts and arthropod vectors, as well as inquiries into the attempts of farmers to mitigate the impact of such diseases and the efforts of scientists to contain them through biomedical research and immunological interventions wherever possible.

Local knowledge about livestock diseases and the hybridization of knowledge are wide-ranging subjects that could be explored in the veterinary context, from a historical as well as an ethnographic perspective. Whereas social histories of disease and biomedicine are now the norm in relation to human medicine, they remain strikingly absent from the veterinary sphere. We know very little, for example, about "indigenous" knowledge or folk practices surrounding popular conceptualizations of infection or how rural communities treated their animals and tried to protect them from the onslaught of disease. Such studies need not be limited to the developing world, where European colonial powers, in particular, constructed a Manichean divide between Western, modern, progressive science on the one hand and the allegedly primitive medical practices of the colonized on the other. As Abigail Woods demonstrates here in her chapter on the attitudes of British farmers to the veterinary profession in the years preceding the World War II, skepticism about the value of science was rife in Europe, too.

Skepticism about the role of science and scientific policies has also been voiced more recently, especially in the wake of the slaughter policy exercised in Great Britain and the Netherlands in 2001 in response to foot-and-mouth disease. But histories of anti-science have a far longer history and have not been greatly explored in the literature. In the nineteenth century, the antivivisectionist movement was particularly strong in Great Britain and possibly elsewhere. Histories of such popular movements constitute an interesting lens through which to assess the relationship among scientists, states, and the wider public. Questions surrounding the value of science and the accountability of scientists to citizens at large, especially in democratic states, remain particularly important in a world in which research into genetic engineering and xenotransplantation raises ethical questions about the nature of existence and the desirability or otherwise of preserving a biological divide between humans and other mammals. Popular responses to modern science and medicine have historical precedents that should be examined for the veterinary sphere.

Finally, far more work could be carried out into the nature of veterinary medicine, not only its epistemology as discussed in the introduction but also the way in which professional exponents of animal health have shored up state bureaucracies and international organizations. Whereas there have been publications about institutions like the London and Liverpool Schools of Tropical Medicine, as well as the World Health Organization,[3] there is a lack of equivalent historical literature on veterinary laboratories and international bodies such as the Organisation Mondiale de la Santé Animale (OIE: World Organization for Animal Health). Such studies would enrich our understanding of the historical development of scientific research in both the national and international contexts and provide insights into the expansion of political and scientific networks, the politics of disease control, and the reasons that specific infections attained relative prominence at particular junctures in time.

These historical topics, among others, are pertinent for trying to comprehend today's world. The historical provenance and logic behind ideas surrounding disease control (be it by slaughter, quarantines, vaccination, or stock dipping), as well as the efficacy of such policies in the field, provide lessons for future livestock management. So too do analyses of popular responses to official interventions, based on the dominant scientific theories of the day. Historically and currently, people have resisted and continue to resist quarantines because they disrupt human and animal movements, damage trade, and potentially ruin individual livelihoods. They have often also proved to be breachable. Stock dipping, too, has been unpopular, as case studies from Africa have revealed.[4] Contemporary opposition to slaughter is no longer limited to farmers' expressions of anger about a loss of assets or disputes about compensation, but now through the media, attracts a broader urban audience that questions the necessity for such actions, especially in the case of diseases like foot-and-mouth for which there is a vaccine. Vaccination has perhaps been the least controversial form of livestock-disease prevention, as long as the inoculations have been safe, effective, accessible, and affordable. However, limits to scientific knowledge mean that prophylaxes are not available for all infections and some vaccinations, such as those against tick-borne diseases, are often unreliable, short-lived, and difficult to administer. There are also no vaccines against toxicoses (plant and fungal poisonings acquired through grazing or ingesting contaminated fodder) or worm infestations that claim millions of livestock lives every year throughout the world. In the absence of universal panaceas for disease control, managing livestock infections will continue to involve a variety of strategies to tackle specific conditions and will result in a mixture of pastoral and public responses. New histories of livestock economies, as well as studies into the shifting cultural conceptualizations of diseases and the varying impact of human efforts to overcome them, will provide useful and interesting contexts for analyzing contemporary veterinary and human-health issues more fully.

Notes

1. Myron Echenberg, *Plague Ports: The Global Urban Impact of Bubonic Plague, 1894–1901* (New York: New York University Press, 2007).

2. Arthur King, A. R. Fooks, M. Aubert, and A. I. Wandeler, eds., *Historical Perspectives of Rabies in Europe and the Mediterranean Basin* (Paris: OIE, 2004); Neil Pemberton and Michael Worboys, *Mad Dogs and Englishmen: Rabies in Britain, 1830–2000* (Basingstoke, England: Palgrave Macmillan, 2007).

3. For example, Paul Weindling, ed., *International Health Organisations and Movements* (Cambridge: Cambridge University Press, 1995); Javed Siddiqi, *World Health and World Politics: The World Health Organisation and the UN System* (London: Hurst, 1995); Lise Wilkinson and Anne Hardy, *Prevention and Cure: The London School of Hygiene and Tropical Medicine; A 20th Century Quest for Global Public Health* (London: Kegan Paul, 2001).

4. For example, Colin Bundy, "'We don't want your rain, we won't dip': Popular Opposition, Collaboration and Social Control in the Anti-Dipping Movement, 1908–16," in *Hidden Struggles in Rural South Africa*, ed. William Beinart and Colin Bundy (London: James Currey, 1987): 191–221.

Livestock Diseases

Anthrax

Caused by spores of the genus *Bacillus anthracis,* anthrax is a zoonosis that affects all types of domestic animals, as well as antelope. It can spread to humans through contact with sick animals and their products. The spores lie in the ground and can remain dormant for a long time. Infection is through contact with a sick animal, through ingestion of spores, and possibly through insect bites. It can now be prevented by vaccination.

Brucellosis (contagious abortion; undulant fever; Malta fever)

Caused by the *Brucella* bacteria, brucellosis was historically a major impediment to increasing the size and yield of herds throughout the world. All types of domestic animals are susceptible to particular strains of this disease, which results in abortion and infertility. The germ is concentrated in aborted fetuses and the uterine fluids of an infected animal. It is also shed in semen and milk. Humans can contract the disease through contact with infected animals, carcasses, or unpasteurized milk. Brucellosis is now preventable by animal vaccination.

Contagious Bovine Pleuropneumonia

Historically, this disease has been present in all parts of the world. It is caused by the *Mycoplasma mycoides* and spread between cattle in water droplets. Symptoms include fever, anorexia, and breathing problems. In serious cases, septicemia sets in, damaging the internal organs. Mortality rates are about 50 percent. Many of the cattle that do survive become carriers. They show no symptoms but are a danger to the rest of the herd. Some countries have managed to eradicate this disease by slaughtering infected herds, imposing strict quarantines, restricting animals' movements, and more recently, through vaccinating herds.

East Coast Fever (Theilerosis)

Also known as African Coast fever, this disease is a major problem for pastoralists in sub-Saharan Africa. It is caused by a parasite, *Theileria parva*, and spread by the brown tick, *Rhipicephalus appendiculatus*. Symptoms include a nasal and lachrymal discharge, as well as anorexia. Mortality is especially high in herds with no prior exposure to the disease. Those cattle that do survive have lifelong immunity but are carriers—hence, a potential danger to the rest of the herd. In the early twentieth century, control was by regular stock dipping to kill the ticks, accompanied by internal quarantines and restrictions on cattle movements. Vaccines now exist but are difficult to store and administer. The brown tick is also responsible for spreading *Theileria lawrenci*, the cause of corridor disease in parts of sub-Saharan Africa.

Foot-and-Mouth Disease (FMD)

This is a very infectious viral disease that affects all types of domestic animals. It is characterized by vesicles on the mouth, teats, and hooves. Although morbidity is up to 100 percent, mortality rates are low. The disease spreads through direct contact with infected animals. Because of international trading agreements, which prize countries with FMD-free status, many countries have resorted to mass culling to eradicate the infection. However, FMD can be effectively controlled through vaccination.

Footrot

Footrot is the result of infection by bacteria that causes lesions to appear on the skin between the claws of sheep. These lesions can deepen so that the horny part of the hoof can become almost completely detached from the rest of the foot. The disease causes lameness, fever, anorexia, loss of condition, and, ultimately, death. Today, the disease is treated with drugs and chemical footbaths. Footrot is common throughout the world, but climate, vegetation, and other environmental factors can influence its distribution.

Fowl Cholera

This is a contagious disease that affects both domestic and wild birds. It is caused by the *Pasteurella multocida* and spread in secretions from the mouth and eyes. Rodents are the prime carriers. The disease can be prevented and controlled through sanitary measures in the hen house, as well as by sulphonamides and antibiotics.

Glanders

A disease of horses, glanders is normally fatal. It is characterized by the development of nodules on the skin as well as in the lungs and other internal organs. It is caused by *Burkholderia mallei* that disseminates through nasal secretions and pus from skin ulcers. Shared water, grazing, and fodder facilitate transmission. Historically, it has affected all continents. There is no vaccine, but there are antibiotic treatments that are deployed in areas where the disease is endemic.

Malignant Catarrhal Fever (snotsiekte)

This is primarily a disease of domestic cattle and water buffalo. Sheep and wildebeest are the main carriers. In Africa, cattle that graze on land where wildebeest have given birth are particularly susceptible to infection as the virus exists in the placenta and vaginal discharges and thus contaminates the grasslands. The disease can also spread between cattle through nasal secretions. There is no known treatment or vaccine.

Nagana (Trypanosomosis in livestock)

Nagana affects all types of livestock and is limited to Africa, where the tsetse fly (*Glossina*) exists. Tsetse pass on the parasites (trypanosomes) from infected to healthy animals. The parasites target the red blood cells, bringing on anemia and eventual death. There is no vaccine, and there were no effective treatments before 1945. However, in West Africa, some varieties of cattle developed a resistance to the trypanosomes, rendering them trypanotolerant. Nevertheless, trypanotolerant cattle will succumb if badly nourished or if tsetse strike is particularly intense.

Newcastle Disease

This is a respiratory disease of birds that is exceedingly virulent and results in high mortality. It is spread by water droplets or by contaminated water and feed. The disease is prevalent in many parts of the developing world, and many countries have import restrictions on the introduction of poultry from infected nations.

Paralytic Rabies in Cattle

Rabies is an acute viral encephalomyelitis that affects mammals. From a livestock perspective, rabies has been particularly detrimental to the cattle economies of some Latin American countries and the Caribbean Islands. The disease is spread by vampire bats (*Desmodus rotundus*). The virus attacks the nervous system, paralyzing the throat muscles, causing excessive salivation. Normally, the animals do not become vicious. Stock owners often mistake the symptoms for an obstruction in the throat and windpipe and may use their bare hands to try to remove it or administer drugs, potentially coming into contact with infective saliva. Preventative inoculation for paralytic rabies has been available for cattle since the 1960s. Vampire bats can also pass on paralytic rabies to humans.

Redwater (Texas fever, Babesiosis)

Redwater is characterized by red urine. Ticks of the genus *Boophilus* spread the disease, which is caused by protozoa, either *Babesia bigemina* or *Babesia bovis*. Herds that have had had a lifetime of exposure to a particular strain acquire some resistance to that form of the disease, so long as they are not malnourished and tick numbers are not too great. In the nineteenth century, farmers dipped their animals in arsenic preparations to kill the ticks. More modern preventatives include vaccines and chemotherapeutic treatments.

Rinderpest (cattle plague)

European accounts often referred to rinderpest as cattle plague, although, historically, a lack of diagnostic tools has meant that other diseases might have come under that same name. The disease affects all cloven-hoofed animals, including domestic cattle, African buffalo, and various species of antelope. The disease was prevalent in Europe and Asia until the nineteenth and twentieth centuries and was introduced to Africa during the period of European colonization in the late nineteenth century. It is caused by a *Morbillivirus*, which is present in the secretions and excretions of infected animals, making it easy for the disease to spread through nasal droplets, as well as through contaminated feed and water. Mortality is very high. Since the twentieth century, it has been possible to control the disease through vaccination, but earlier policies involved slaughtering infected and in-contact animals, as well as quarantines and restrictions on the movement of cattle.

Sheep Pox

Historically, this disease has broken out in many parts of Eurasia and Africa. It causes skin eruptions and is probably transmitted by air. Today, it can be prevented with vaccines.

Sheep Scab (Psoroptic mange)

Mites bore into the skin and cause this disease, which is characterized by the appearance of large scaly lesions on the woolly parts of sheep. This reduces the value of the wool and can lead to a decline in milk and meat yields. In the nineteenth century, farmers developed dips to kill the mites. Dipping and the administration of drugs such as ivermectin are the main methods for controlling the disease today.

Surra

Surra is the Indian name for a form of trypanosomosis that affects horses and camels. Biting tabanid flies spread the disease, which is caused by the parasite *Trypanosoma evansi*. As in nagana, the parasites destroy the red blood cells, bringing on anemia and the wasting away of the animal. The disease is limited in distribution to Asia, the Middle East, and North Africa. There are no vaccines.

Swine Fever

Classical and African swine fever are febrile diseases in pigs that can result in devastating epidemics. Historically, all continents have been infected. Control has been by slaughter, although there is now a vaccine.

Tuberculosis (TB)

This disease affects all vertebrates, and bovine TB is a major problem globally. The usual route of infection is from droplets expelled from the lungs of an ailing animal. Humans can acquire the disease through contaminated milk, which from the late nineteenth century, led to the gradual spread of pasteurization as a means of sterilizing milk. In countries where eradication is feasible, herds are normally slaughtered. Elsewhere, sick animals are segregated and may be treated with drug therapies. There is no vaccine for TB in animals.

Select Bibliography

Anderson, David. *Eroding the Commons: The Politics of Ecology in Baringo, Kenya 1890–1963.* Oxford: James Currey, 2002.

Andreas, Christian. "The Lungsickness Epizootic of 1853–1857: An Analysis of Its Socio-Economic Impact and the Ensuing Reactions in the Cape Colony and Xhosaland." Master's thesis, University of Hannover, 2003.

Bankoff, Greg. "A Question of Breeding: Zootechny and Colonial Attitudes towards the Tropical Environment in the Late Nineteenth-Century Philippines." *Journal of Asian Studies* 60, no. 2 (2001): 413–38.

Bankoff, Greg, and Sandra Swart, eds. *Breeds of Empire: The Invention of the Horse in Southeast Asia and Southern Africa, 1500–1950.* Copenhagen: NIAS Press, 2007.

Beinart, William. *The Rise of Conservation in South Africa: Settlers, Livestock and the Environment, 1770–1950.* Oxford: Oxford University Press, 2003.

———. "Vets, Viruses and Environmentalism at the Cape." *Paideuma* 43 (1997): 227–52.

Bierer, B. W. *A Short History of Veterinary Medicine.* East Lansing: Michigan State University Press, 1955.

Blancou, Jean. *History of the Surveillance and Control of Transmissible Animal Diseases.* Paris: Office International des Épizooties, 2003.

Boomgaard, Peter, and David Henley, eds. *Smallholders and Stockbreeders: Histories of Foodcrop and Livestock Farming in Southeast Asia.* Leiden: KITLV Press, 2004.

Brassley, Paul. "Output and Technical Change in Twentieth-Century British Agriculture." *Agricultural History Review* 48, no. 1 (2000): 60–84.

Brock, Thomas D. *Robert Koch: A Life in Medicine and Bacteriology.* Madison, WI: Science Tech Publishers, 1988.

Brown, Karen. "From Ubombo to Mkhuzi: Disease, Colonial Science and the Control of Nagana (Livestock Trypanosomosis) in Zululand, South Africa,

c. 1894–1953." *Journal of the History of Medicine and Allied Sciences* 63, no. 3 (2008): 285–322.

———. "Poisonous Plants, Pastoral Knowledge and Perceptions of Environmental Change in South Africa, c. 1880–1940." *Environment and History* 13, no. 3 (2007): 307–32.

———. "Tropical Medicine and Animal Diseases: Onderstepoort and the Development of Veterinary Science in South Africa, 1908–1950." *Journal of Southern African Studies* 31, no. 3 (2005): 413–529.

Brown, Karen, and Daniel Gilfoyle, eds. "Livestock Diseases and Veterinary Science." Special issue, *South African Historical Journal* 58 (2007): 1–141.

Clutton-Brock, Juliet. *Horse Power: A History of the Horse and the Donkey in Human Societies.* Cambridge, MA: Harvard University Press, 1992.

Cranefield, Paul F. *Science and Empire: East Coast Fever in Rhodesia and the Transvaal.* Cambridge: Cambridge University Press, 1991.

Crosby, Alfred. *The Columbian Exchange: Biological and Cultural Consequences of 1492.* Westport, CT: Greenwood, 1972.

Cunningham, Andrew, and Perry Williams, eds. *The Laboratory Revolution in Medicine.* Oxford: Oxford University Press, 1992.

Davis, Diana K. "Desert 'Wastes' of the Maghreb: Desertification Narratives in French Colonial Environmental History in North Africa." *Cultural Geographies* 11, no. 4 (2004): 359–87.

———. "Prescribing Progress: French Veterinary Medicine in the Service of Empire." *Veterinary Heritage* 29, no. 1 (2006): 1–7.

Fisher, John. "Cattle Plagues Past and Present: The Mystery of Mad Cow Disease." *Journal of Contemporary History* 33, no. 2 (1998): 215–28.

———. "The Economic Effects of Cattle Disease in Britain and Its Containment, 1860–1900." *Agricultural History* 54, no. 2 (1980): 278–94.

———. "Not Quite a Profession: The Aspirations of Veterinary Surgeons in England in the Mid-Nineteenth Century." *Historical Research* 66, no. 161 (1993): 284–302.

———. "A Pandemic (Panzootic) of Pleuropneumonia, 1840–1860." *Historica Medicinae Veterinariae* 11, no. 1 (1986): 26–32.

———. "To Kill or Not to Kill: The Eradication of Contagious Bovine Pleuropneumonia in Western Europe." *Medical History* 47, no. 3 (2003): 314–31.

Ford, John. *The Role of Trypanosomiasis in African Ecology: A Study of the Tsetse Fly Problem.* Oxford: Clarendon, 1971.

Geison, Gerald L. *The Private Science of Louis Pasteur.* Princeton, NJ: Princeton University Press, 1995.

Giblin, James. "East Coast Fever in Socio-historic Context: A Case Study from Tanzania." *International Journal of African Historical Studies* 23, no. 3 (1990): 401–21.

———. "Trypanosomiasis Control in African History: An Evaded Issue?" *Journal of African History* 31, no. 1 (1990): 59–80.

Gilfoyle, Daniel. "Veterinary Research and the African Rinderpest Epizootic: The Cape Colony, 1896–98." *Journal of Southern African Studies* 29, no. 1 (2003): 133–54.

———. "Veterinary Science and Public Policy at the Cape, 1877–1910." DPhil. thesis, University of Oxford, 2002.

Hardy, Anne. "Pioneers in the Victorian Provinces: Veterinarians, Public Health and Urban Animal Economy." *Urban History* 29, no. 3 (2002): 372–87.

———. "Professional Advantage and Public Health: British Veterinarians and State Veterinary Services, 1865–1939." *Twentieth Century History* 14, no. 1 (2003): 1–23.

Hughes, Lotte. *Moving the Maasai: A Colonial Misadventure.* Basingstoke, England: Palgrave MacMillan, 2006.

Huygelen, Constant. "The Immunization of Cattle against Rinderpest in Eighteenth-Century Europe." *Medical History* 41, no. 2 (1997): 182–96.

Jones, Susan. *Valuing Animals: Veterinarians and Their Patients in Modern America.* Baltimore, MD: Johns Hopkins University Press, 2003.

Kjekshus, Helge. *Ecology Control and Economic Development in East African History: The Case of Tanganyika, 1850–1950.* London: James Currey, 1996.

Kohler, Robert. *Landscapes and Labscapes.* Chicago: University of Chicago Press, 2002.

Koolmees, Peter. "The Role of Veterinary Medicine in the Development of Factory Farming." In *The Human-Animal Relationship: Forever and a Day,* edited by Francien de Jonge and Ruud van den Bos, 249–64. Assen, Netherlands: Van Gorcum, 2005.

Leach, Melissa, and Robin Mearns, eds. *The Lie of the Land: Challenging Received Wisdom on the African Landscape.* Oxford: James Currey, 1996.

MacOwan, K. D. S. "The Development of a Livestock Industry in Kenya." *Veterinary History* 8, nos. 1–2 (1994): 29–37.

Martin, John E. *The Development of Modern Agriculture: British Farming since 1931.* Basingstoke, England: Macmillan, 2000.

———. *A Legacy and a Promise: The First One Hundred Years of the School of Veterinary Medicine.* Philadelphia: University of Pennsylvania, 1984.

Mayhew, Madeleine. "The 1930s Nutrition Controversy." *Journal of Contemporary History* 23, no. 3 (1988): 445–64.

McCracken, John. "Experts and Amateurs: Tsetse, Nagana and Sleeping Sickness in East and Central Africa." In *Imperialism and the Natural World,* edited by John MacKenzie, 187–212. Manchester: Manchester University Press, 1990.

McKee, Francis. "The Popularisation of Milk as a Beverage during the 1930s." In *Nutrition in Britain: Science, Scientists and Politics in the Twentieth Century*, edited by David Smith, 50–58. London: Routledge, 1997.

McShane, Clay, and Joel Tarr. "The Centrality of the Horse in the Nineteenth-Century American City." In *The Making of Urban America*, edited by Raymond Mohl, 105–30. Wilmington, DE: SR Books, 1997.

Metivier, Hugh. *A History of the Overseas Veterinary Services, Part II: Trinidad and Tobago, 1879–1958*. London: British Overseas Veterinary Association, 1973.

Milton, Shaun. "Western Veterinary Medicine in Colonial Africa: A Survey, 1902–1963." *Argos* 18 (1998): 313–22.

Mylrea, Peter. *In the Service of Agriculture: A Centennial History of the New South Wales Department of Agriculture*. Sydney: NSW Agriculture and Fisheries, 1990.

Newton, Leslie, and Ronald Norris. *Clearing a Continent: The Eradication of Contagious Bovine Pleuropneumonia from Australia*. Melbourne: CSIRO, 2000.

Nicolaisen, J. *Ecology and Culture of the Pastoral Tuareg: With Particular Reference to the Tuareg of Ahaggar and Ayr*. Copenhagen: National Museum of Copenhagen, 1963.

Pattison, Iain. *The British Veterinary Profession, 1791–1948*. London: J. A. Allen, 1983.

Pemberton, Rita. "The Evolution of Agricultural Policy in Trinidad and Tobago, 1890–1945." PhD thesis, University of the West Indies, 1996.

Perren, Richard. *Agriculture in Depression, 1870–1940*. Cambridge: Cambridge University Press, 1995.

Phimister, Ian. "Meat and Monopolies: Beef Cattle in Southern Rhodesia, 1890–1938." *Journal of African History* 19, no. 3 (1978): 391–414.

Phoofolo, Pule. "Epidemics and Revolutions: The Rinderpest Epizootic in Late Nineteenth-Century Southern Africa." *Past and Present* no. 138 (February 1993): 112–43.

Raby, Geoff. *Making Rural Australia: An Economic History of Technical and Institutional Creativity, 1788–1860*. Melbourne: Oxford University Press, 1996.

Raikes, Peter. *Livestock Policy and Development in East Africa*. Uppsala: NAI, 1981.

Russell, Nicholas. *Like Engend'ring Like: Heredity and Animal Breeding in Early Modern England*. Cambridge: Cambridge University Press, 1986.

Schönherr, W. "History of Veterinary Public Health in Europe in the 19th Century." *Revue scientifique et technique/Office international des epizooties* 10, no. 4 (1991): 985–94.

Smithcors, J. F. *The American Veterinary Profession: Its Background and Development*. Ames: Iowa State University Press, 1963.

Spinage, Clive A. *Cattle Plague: A History.* New York: Kluwer Academia, 2003.

Stalheim, Ole. *The Winning of Animal Health: 100 years of Veterinary Medicine.* Ames: Iowa State University Press, 1994.

Swabe, Joanna. *Animals, Disease and Human Society: Human-Animal Relations and the Rise of Veterinary Medicine.* London: Routledge, 1999.

Tomes, Nancy. *The Gospel of Germs.* Cambridge, MA: Harvard University Press, 1998.

Van Onselen, Charles. "Reactions to Rinderpest in Southern Africa, 1896–97." *Journal of African History* 13, no. 3 (1972): 473–88.

Waddington, Keir. "To Stamp Out 'So Terrible a Malady': Bovine Tuberculosis and Tuberculin Testing in Britain, 1890–1939." *Medical History* 48, no. 1 (2004): 29–48.

———. "'Unfit for Human Consumption': Tuberculosis and the Problem of Infected Meat in Late Victorian Britain." *Bulletin of the History of Medicine* 77, no. 3 (2003): 636–61.

Waller, Richard. "'Clean' and 'Dirty': Cattle Disease and Control Policy in Colonial Kenya, 1900–40." *Journal of African History* 45, no. 1 (2004): 45–80.

———. "Tsetse Fly in Western Narok, Kenya." *Journal of African History* 31, no. 1 (1990): 81–101.

Watts, David. *The West Indies: Patterns of Development, Culture and Environmental Change since 1492.* Cambridge: Cambridge University Press, 1987.

West, G. P., ed. *A History of the Overseas Veterinary Service, Part I.* London: British Veterinary Association, 1961.

Wilkinson, Lise. *Animals and Disease: An Introduction to the History of Comparative Medicine.* Cambridge: Cambridge University Press, 2005.

———. "Rinderpest and Mainstream Infectious Disease Concepts in the Eighteenth Century." *Medical History* 28, no. 2 (1984): 129–50.

Wilson, R. T. "The Incidence and Control of Livestock Diseases in Anglo-Egyptian Sudan, 1916–56." *International Journal of African Historical Studies* 12, no. 1 (1979): 62–82.

Woods, Abigail. "'Flames and Fear on the Farms': Controlling Foot and Mouth Disease in Britain, 1892–2001." *Historical Research* 77, no. 198 (2004): 520–42.

———. *A Manufactured Plague: The History of Foot and Mouth Disease in Britain.* London: Earthscan, 2004.

Worboys, Michael. "Germ Theories and British Veterinary Medicine, 1860–1890." *Medical History* 35, no. 3 (1991): 308–27.

———. *Spreading Germs: Disease Theories and Medical Practice in Britain, 1865–1900.* Cambridge: Cambridge University Press, 2000.

Contributors

David Anderson is professor of African politics and director of the African Studies Centre at the University of Oxford, as well as a fellow of St. Cross College. He has written extensively on the environmental and political history of eastern Africa. His most recent books are *The Khat Controversy: Stimulating the Debate on Drugs, Histories of the Hanged: Britain's Dirty War in Kenya and the End of Empire,* and the forthcoming *Uncivil Society: Violence and Politics in Kenya.* He is currently engaged in a research project on environmental change, settlement, and mobility in the Omo Valley of southwestern Ethiopia.

Martine Barwegen studied animal sciences at Wageningen University in the Netherlands and completed her PhD on the history of animal husbandry in Java. Since then she has worked as an agricultural journalist. Her major publications include her PhD thesis, "Gouden Hoorns: De geschiedenis van de veehouderij op Java, 1850–2000"; "Browsing in Livestock History: Large Herbivores and the Environment in Java, 1850–2000" in the collection *Smallholders and Stockbreeders: Histories of Foodcrop and Livestock Farming in Southeast Asia;* and "De Burgerlijke Veeartsenijkundige Dienst en de uitbraak van de veepest in 1878 op Java," in the journal *Argos.*

Karen Brown is an Economic and Social Research Council fellow at the Wellcome Unit for the History of Medicine, University of Oxford. She is interested in environmental, veterinary, and medical history with a research focus on South Africa. She has published on a number of environmental issues and has just completed a book on the history of rabies in southern Africa. Other recent publications include "Veterinary Entomology, Colonial Science and the Challenge of Tick-borne Diseases in South Africa during the late Nineteenth and Early Twentieth Centuries" in the journal *Parassitologia;* "From Ubombo to Mkhuzi: Disease, Colonial Science and the

Control of *Nagana* (Livestock Trypanosomosis) in Zululand, South Africa, c1894–1955" in the *Journal of the History of Medicine and Allied Sciences;* "Poisonous Plants, Pastoral Knowledge and Perceptions of Environmental Change in South Africa, c. 1880–1940" in the journal *Environment and History;* and "Tropical Medicine and Animal Diseases: Onderstepoort and the Development of Veterinary Science in South Africa 1908–1950" in the *Journal of Southern African Studies.*

William G. Clarence-Smith is professor of the economic history of Asia and Africa at the School of Oriental and African Studies, University of London, and chief editor of the *Journal of Global History.* His latest book is *Islam and the Abolition of Slavery.* His thesis and first book, *Slaves, Capitalists and Peasants in Southern Angola, 1840–1926,* considered the pastoral economy of this region. He has since published on the trade in equids in the Indian Ocean, specifically in Indonesia, and on the raising of horses in Southeast Asia. He is preparing a book on equids and elephants in this region.

Daniel F. Doeppers is emeritus professor of geography and Southeast Asian studies at the University of Wisconsin–Madison. He is currently finishing a book-length study, *Feeding Manila in Peace and War, 1850–1945.* His more recent works include *Population and History: The Demographic Origins of the Modern Philippines* (coedited with Peter Xenos); "Lighting a Fire: Home Fuel in Manila, 1850–1945" in the journal *Philippine Studies;* "Beef Consumption and Regional Cattle Husbandry Systems in the Philippines, 1850–1930s" in the collection *Smallholders and Stockbreeders; Histories of Foodcrop and Livestock Farming in Southeast Asia;* and "Metropolitan Manila in the Great Depression: Crisis for Whom?" in the collection *Capitalism in Asia.*

John Fisher is loosely attached to the Faculty of Business and Law at the University of Newcastle, NSW, Australia. As he spends considerable periods in the United Kingdom, his research interests extend from veterinary history in Australasia to local history in Nottinghamshire. Key recent publications include "The Origins, Spread and Disappearance of Contagious Bovine Pleuro-pneumonia in New Zealand" in the *Australian Veterinary Journal* and "Rochdale Co-operation in New South Wales: A Failure of Cultural Transmission?" in the collection *Shop Till You Drop: Australian Essays on Consuming and Dying.*

Daniel Gilfoyle studied history at the University of the Witwatersrand in Johannesburg and at Birkbeck College, University of London. He completed the degree of doctor of philosophy at St. Antony's College, University of Oxford, and a postdoctoral fellowship at the Wellcome Unit for the History of Medicine, University of Oxford. He has published a number of articles on the history of veterinary science and now works at the National Archives in London.

Ann N. Greene is in the Department of History and Sociology of Science at the University of Pennsylvania in Philadelphia. She is the author of *Horses at Work: Harnessing Power in Industrial America*. Her current research interests are in environmental history and in the history and politics of veterinary medicine and animal sciences in the United States during the late nineteenth and early twentieth centuries.

Lotte Hughes is a lecturer in African arts and cultures at the Ferguson Centre for African and Asian Studies at the Open University and is also a member of the History Department. She works mainly on empire and postcolonial issues in Kenya. Her current research interests focus upon heritage, memory, and identity. She coauthored *Environment and Empire* (with William Beinart) and published her thesis as *Moving the Maasai: A Colonial Misadventure*.

Dominik Hünniger is the academic coordinator at the Lichtenberg-Kolleg (Institute for Advanced Study), University of Göttingen, Germany. He is interested in rural and environmental history, the social history of medicine and the history of knowledge, with a particular focus on early modern Europe. Publications include *Beten, Sammeln, Impfen: Zur Schädlings- und Viehseuchenbekämpfung in der Frühen Neuzeit* (coedited with Katharina Engelken and Steffi Windelen); "Global denken—lokal forschen: Auf der Suche nach dem 'kulturellen Dreh' in der Umweltgeschichte; Ein Literaturbericht" in the journal *WerkstattGeschichte* (coauthored with Richard Hölzl); and the forthcoming "Cattle Plague in Early Modern Germany: Environment and Economy/Knowledge and Power in a Time of Crisis" in the collection *Elements—Continents: Approaches to Determinants of Environmental History and their Reifications*.

Peter A. Koolmees joined the Faculty of Veterinary Medicine, Utrecht University, in 1973, where he studied zoology and history. His thesis was "Symbolen van openbare hygiëne: Gemeentelijke slachthuizen in Nederland, 1795–1940." He now holds a chair in veterinary history at Utrecht and is editor of *Geschiedenis der Geneeskunde* and *Argos*. He was president of the World Association for the History of Veterinary Medicine from 2000 until 2004. Recent publications include "The Development of Veterinary Public Health in Western Europe, 1840–1940" in the journal *Sartoniana;* "The Role of Veterinary Medicine in the Development of Factory Farming" in the collection *The Human-Animal Relationship: Forever and a Day;* and "From Stable to Table: The Development of the Meat Industry in the Netherlands, 1850–1990" in the collection *Exploring the Food Chain: Food Production and Food Processing in Western Europe, 1850–1990.*

Saverio Krätli is a freelance researcher based in the United Kingdom and specializing in the interface between science and policy, particularly in pastoral production strategies and education. He has carried out fieldwork among pastoralists in East and West Africa (Uganda, Kenya, and Niger), with the Institute of Development Studies (IDS, University of Sussex), and with the International Institute for Environment and Development (IIED, London). He is also editor of the journal *Nomadic Peoples.*

Robert Peden spent twenty-five years working on sheep stations (ranches) in the South Island high country before returning to university in 1999. He has a master's degree in history from the University of Canterbury and a PhD from the University of Otago. He was the 2009 Claude McCarthy Fellow, attached to the History Department of the University of Canterbury, and is completing a book about pastoralism and the transformation of the tussock grasslands (rangelands) of the South Island of New Zealand, 1841 to 1912.

Rita Pemberton is a senior lecturer and head of the Department of History at the St. Augustine campus of the University of the West Indies. Her research centers on the history of health and environment in the Caribbean and the history of Tobago. Her recent publications include "Shaping the Caribbean Environment: The Impact of India" in the collection *Situating Environmental History;* "A Centre in the Periphery: His Majesty's Botanic Garden, St. Vincent, 1785–1815" in the collection *Beyond Tradition: Reinterpreting the Caribbean Historical Experience;* "Water and Related Issues in 19th Century Trinidad" in the *Journal of Caribbean History;* and "Ports and Health: Port of Spain, 1838–1900" in the collection *El Golfo-Caribe y sus puertos,* vol. 2, *1850–1915.*

Robert John Perrins is dean of the Faculty of Arts at Acadia University in Nova Scotia, Canada, where he is also a professor in the Department of History and Classics specializing on the histories of medicine, disease, and East Asia. He is currently working on the history of the development of colonial public health in Manchuria during the early twentieth century and on the history of plague in modern Asia.

Abigail Woods is a lecturer in the history of medicine at Imperial College, London. She trained as a veterinary surgeon at Cambridge University before completing an MSc and PhD in the history of medicine at Manchester University. Research interests encompass the history of livestock disease, preventive veterinary medicine, animal health policy, and the British veterinary profession. Her book, *A Manufactured Plague: The History of Foot and Mouth Disease in Britain, 1839–2001,* was published in 2004.

Index

Rift Valley, 147, 150–52, 253, 255, 257–58, 262
rinderpest, 2–4, 7–8, 20–28, 37, 76, 92,
 97–100, 102–3, 108–25, 152, 154–55,
 157–58, 185, 196, 199, 207–8, 262,
 269–70, 278. *See also* cattle plague
Royal Veterinary College, London, 190
Russia, 21

Sahel, 12, 131, 235, 237, 239, 242
sanitary measures, 11, 277
scab, 3, 13, 183–89, 191, 217, 279
Scab Acts (Australia), 184–85, 187, 189
Scab Board (Australia), 187, 191
Scab Branches (Australia), 181, 183,
 187–90
Schleswig-Holstein, 6, 8, 77, 79, 81,
 83–84, 86–87, 90
scrub stock, 251, 255, 262
sedentary farmers, 12, 238–42
settler beef industry, 251, 256
settler economies, 3, 12, 254
settler farmers, 11, 12, 150, 259, 262
settlers, 7, 10, 147, 150, 155, 158, 181–82,
 184, 198, 204–5, 207, 218, 251–55, 258,
 262–64
settler stock, 150, 154, 256
sheep, 3, 7, 13, 22, 25, 30–31, 45, 99, 108,
 111–12, 157, 163, 165, 180–88, 200–207,
 215–29, 262–63, 271, 276–77, 279
sheep pox, 9, 196, 207, 210, 279
sheep stations, 218–20, 222, 225–26
Sino-Japanese War, 206
skins, 79, 83–84, 237
slaughterhouses. *See* abattoirs
Smith, Horatio, 46–48, 50–51
Soriano, Ramon, 118–19
South Africa, 2, 4, 7, 9, 11, 13, 22, 135, 221,
 228, 251–52
South America. *See* Latin America
South Manchuria Railway Company.
 See Mantetsu (South Manchuria
 Railway Company)
South Manchuria Veterinary
 Association, 195, 197, 208
Spain, 109, 112
stamping out. *See* cull-and-slaughter
 policy

State Veterinary Research Institute,
 Rotterdam, 30
State Veterinary School, Utrecht, 27–28
Steele-Bodger, Harry, 62, 64, 67
sterility. *See* infertility
Stockdale, Frank, 254, 257, 259
stock inspectors, 155–56, 190
Stock Owners Association (Kenya), 258,
 260, 262
Sumatra, 94, 97, 100, 110, 124, 130–31
surra, 3, 9, 96, 103, 130–33, 136, 139, 279
swine fever, 19, 28–31, 33–34, 36–37,
 173–74, 279

Taiwan, 117, 197–98
tallow, 84, 86, 223
Tanganyika, 258–60, 264
tarbagan, 199
tetanus, 5, 134
Texas fever (redwater), 4, 46, 52, 133,
 154, 278
Thailand, 109, 131, 135–37
ticks, 2, 4, 10–11, 133, 150, 152–54, 156–57,
 160, 167, 276, 278
tourism, 33–34
toxicoses. *See* plant poisonings
 (toxicoses)
trade, 3–4, 6–8, 13, 19–21, 24, 33, 36–37,
 46, 53, 79–83, 85–86, 99, 104, 109–12,
 117, 124, 131, 181, 187, 216, 225, 227–28,
 253, 256, 259, 262–64, 270, 273
trade barriers, 24, 112
trade embargoes, 8, 22–23
transhumance, 12, 235, 239
Trinidad and Tobago, 12, 164–75, 270–71
Trinidad and Tobago Bureau of Animal
 Industry, 170
tropical diseases, 3–4, 9, 103, 129, 133
tropical veterinary medicine, 3
tropics, 9, 109, 130
trypanosomosis (*nagana*), 2–3, 9–10,
 96, 130, 132, 277, 279
trypanotolerance, 132
tsetse fly, 2, 9–10, 96, 131, 150, 152, 157,
 277
Tuareg, 233, 238–39, 242, 246
tuberculin testing, 63, 167, 170